Bounded Rationality and Politics

# THE AARON WILDAVSKY FORUM FOR PUBLIC POLICY

*Edited by Lee Friedman*

This series is to sustain the intellectual excitement that Aaron Wildavsky created for scholars of public policy everywhere. The ideas in each volume are initially presented and discussed at a public lecture and forum held at the University of California.

## AARON WILDAVSKY, 1930–1993

"Your prolific pen has brought real politics to the study of budgeting, to the analysis of myriad public policies, and to the discovery of the values underlying the political cultures by which peoples live. You have improved every institution with which you have been associated, notably Berkeley's Graduate School of Public Policy, which as Founding Dean you quickened with your restless innovative energy. Advocate of freedom, mentor to policy analysts everywhere."

> *(Yale University, May 1993, from text granting the honorary degree of Doctor of Social Science)*

# Bounded Rationality and Politics

Jonathan Bendor

UNIVERSITY OF CALIFORNIA PRESS
*Berkeley* · *Los Angeles* · *London*

University of California Press, one of the most distinguished
university presses in the United States, enriches lives around
the world by advancing scholarship in the humanities, social
sciences, and natural sciences. Its activities are supported
by the UC Press Foundation and by philanthropic
contributions from individuals and institutions. For more
information, visit www.ucpress.edu.

University of California Press
Berkeley and Los Angeles, California

University of California Press, Ltd.
London, England

Library of Congress Cataloging-in-Publication Data

Bounded rationality and politics / by Jonathan Bendor ... [et al.].
   p.   cm.—(The Aaron Wildavsky forum for public policy; 6)
Includes bibliographical references and index.
ISBN 978-0-520-25946-1 (cloth : alk. paper)
ISBN 978-0-520-25947-8 (pbk. : alk. paper)
   1. Decision making—Political aspects.   2. Simon, Herbert A.
(Herbert Alexander), 1916–2001.     3. Organizational behavior—
Political aspects.     4. Social sciences—Philosophy.     I. Bendor,
Jonathan B.
   BF441.B645   2010
   320.01′9—dc22                                   2009052478


19   18   17   16   15   14   13   12   11   10
10   9   8   7   6   5   4   3   2   1

*For Linda,*
*who's understood what it's meant*

# Contents

# Figures

# Preface

I have been studying the relation between bounded rationality and politics off and on for almost thirty years. (I work slowly.) The essays in this book, however, are all from the last dozen or so years. All of the chapters share a simple premise: cognitive constraints often affect judgment and choice in powerful ways. Hence, models of full (unbounded) rationality are ignoring causally important variables. How much they are missing depends on the difficulty of the problems tackled by decision makers. The harder their problems, the more cognitive constraints bind (Simon 1996) and the greater their causal impact.

But for several reasons this book is not another critique of rational choice (RC) theories. First, I am more interested in providing alternatives to classical models of decision making than in critiquing them. Of course the two activities are intertwined, but the latter badly needs the former: you can't beat something with nothing. As Shepsle has cautioned us, "The First Law of Wing Walking [is] don't let go of something unless you have something else to hang onto" (1996, p. 217). This advice is based on a sound appreciation of the sociology of science, and a good case can be made for it as a normative decision rule as well.

Second, I think that most critics of RC theories (e.g., Green and Shapiro 1994) have seriously underestimated the contributions that many of these theories have made to political science. They have given us insights into overlooked problems such as those of collective action; they have given us bold predictions such as the median voter theorem. Critics point

out that RC's bold predictions often fall short empirically. But the point is
not to be always correct. That's impossible; striving for perfection produces
flabby theories that can't be falsified. The point, per Popper (1963), is to
discover and correct one's errors as rapidly as possible. There's no shame
in being wrong. Much better to be "strong and wrong" (Schotter 2006), to
have produced "a beautiful theory [murdered] by a gang of brutal facts" (La
Rochefoucauld [1678]) than one that makes no interesting claims at all.[1]

RC theorizing has also given our discipline a badly needed measure
of intellectual coherence. This is important not only for research; it aids
teaching as well. For example, I think that a wise pedagogical strategy
for doctoral students is to study RC theories first. Only after they have a
good grasp of this research program should they plunge into the less well-
structured bounded rationality program. People who teach introductory
game theory always begin with models of complete information, which are
much easier to understand than those with incomplete information. In the
context of this book, the point is reflexive: if we take seriously cognitive
constraints—including our own and our students'—then it's easy to see why
the models of complete rationality are intellectually important benchmarks
and starting places.

For all these reasons, these essays should be taken as providing friendly
competition to RC models—but competition nonetheless. My decision to
collect these essays into a book was prompted by a request from the University of California, Berkeley's Graduate School of Public Policy to give
the 2004 lecture honoring the memory of Aaron Wildavsky. (The GSPP
typically asks the lecturer to provide a book manuscript as well as the
lecture.) Because Aaron was one of my advisors and had a major impact
on my thinking, I was happy to give the lecture. The book, however, was
another matter. I hadn't written a book in twenty years and had been
focusing instead on articles. Then Nelson Polsby stepped in and gave me
good advice (as he so often did throughout my career): use the Wildavsky
lecture as chapter 1, insert related and already written articles as the heart
of the book, and write a new essay as a conclusion.[2] Done!

Due to this genesis, parts of chapter 1 have a personal tone. Aaron meant
a lot to me, and the GSPP lecture reflected that. Because it is good to honor
our teachers in print, I have retained that tone here.

This personal tone recurs in parts of the last chapter. There I pay my
respects to my main advisor and thesis chair, Martin Landau, as well
as to Aaron. Marty passed away in December 2004, so he was on my
mind when I started thinking about this book. His influence on me was
huge. For a commemoration honoring Marty, I wrote a brief essay on the

organization-theoretic foundations of Marty's philosophy of science, a combination that is central to his intellectual legacy. Because these ideas address how scientific procedures and institutions can ameliorate the cognitive constraints of individual scientists and because the general topic of organizational compensation for the bounded rationality of individuals is (not coincidentally!) the topic of this book's last chapter, most of the intellectual eulogy for Marty is included in chapter 7.[3] Someday I'd love to build on these ideas and hypotheses of Marty's; for now a sketch must do.

Because I find it impossible to refer to people who meant much to me by their last names, I refer to both of my teachers by their first names. (Evidently Marty's Army service in World War II made him allergic to being called "Landau." I can't do it.) With only one exception, the other five chapters use the normal aseptic academic style. That exception is a brief passage in chapter 2 about Herbert Simon's contributions to political science. Though I never took a class from him—indeed, I met him only twice and had only one sustained conversation with him—I learned more from Simon than from anyone else. How could it be otherwise? I still recall (at least I think I do; long-term memories are so fallible!) reading chapter 6 of *Organizations*, on "cognitive limits...," as a junior at Berkeley. (I even have a memory of the study room where I read it.) That chapter's influence on me was decisive. I never stopped reading Simon's work. The man was a genius.

In sum, this book is a tribute to Aaron Wildavsky, Marty Landau, and Herbert Simon.

Having already described chapters 1, 2, and 7, let me finish the outline. Chapters 3, 4, and 5 are the intellectual core of the book. They study a concept central to the BR program, that of *heuristics*. Chapter 3 briefly analyzes the program's most famous heuristic: satisficing. Chapter 4 formalizes several others, including incremental search, put forward by Lindblom (1959, 1965) and Braybrooke and Lindblom (1963). Chapter 5 studies a putting-out-the-fire heuristic. This chapter's substantive context, organizational reliability, was inspired by Marty's pathbreaking article on this topic (Landau 1969); the analysis of the damage done by unrealistically high aspirations was strongly influenced by Aaron's work on effective policy analysis (Wildavsky 1979).

Chapter 6 criticizes a line of work, garbage can theory, that is often associated with the BR program. My coauthors, Terry Moe and Ken Shotts, and I argue that this association constitutes a misunderstanding of both the larger program and garbage can theory. An addendum to the chapter, recently coauthored with Ken Shotts, contains a spare new model of garbage can processes.

I have benefited enormously from working with wonderful coauthors. Working with these friends has constituted an organizational solution to my cognitive constraints.[4] It has been more than fun: it's been one of the chief joys of my professional life.

Chapter 3 is based on a long-standing project on satisficing with Sunil Kumar and David Siegel. Our coauthored paper has gone through many incarnations; chapter 3 is a fragment of one of these. Chapter 5 was coauthored with Sunil; chapter 6, with Terry Moe and Ken Shotts. Part of chapter 1 is based on "The Empirical Content of Adaptive Models," coauthored with Daniel Diermeier and Michael Ting. Part of chapter 7 is from "Rethinking Allison's Models," which Tom Hammond and I wrote.

Although our joint work does not appear in this volume, I also thank Dilip Mookherjee, Debraj Ray, and Piotr Swistak, and (most recently) Nathan Collins: these friends and I have worked on models of bounded rationality over the last fifteen years. (Dilip and I go back even further, but our early papers used full-rationality assumptions.)

I thank the following scholars for their helpful comments on earlier versions of the papers included in this book: Thom Baguley, Bill Barnett, Dave Baron, Jose Bermudez, Dave Brady, Brandice Canes-Wrone, Kathleen Carley, Dan Carpenter, Glenn Carroll, Kelly Chang, John Conlisk, Robyn Dawes, Mandeep Dhami, Persi Diaconis, Baruch Fischhoff, Elizabeth Gerber, Itzhak Gilboa, Dan Goldstein, Mark Green, John Jost, Daniel Kahneman, Rod Kramer, Krish Ladha, David Laitin, Martin Landau, Susanne Lohmann, Bill Lovejoy, Duncan Luce, Howard Margolis, William McKelvey, John Meyer, Elijah Millgram, Lawrence Mohr, Jeff Moore, Barry Nalebuff, Joe Oppenheimer, John Padgett, Paul Pfleiderer, Joel Podolny, Nelson Polsby, Bob Powell, Tonya Putnam, John Scholz, Ken Shotts, Jay Silver, Neil Smelser, Art Stinchcombe, Serge Taylor, Philip Tetlock, Richard Thaler, Fred Thompson, Mike Ting, John Wagner III, Jin Whang, and David Yoffie.

I am especially grateful to David Braybrooke, Daniel Diermeier, Dale Griffin, Tom Hammond, Bernardo Huberman, Charles Lindblom, Herbert Simon, Joel Sobel, and Piotr Swistak for their insightful comments on several papers/chapters. (Tom wins the prize for diligence: he read and critiqued four papers.)

Faten Sabry and Abe Wu did numerical simulations that helped guide the original Bendor-Moe-Shotts exploration of the Cohen-March-Olsen formulation. Chris Stanton's swift computational work on the new Bendor-Shotts model of garbage can processes was very valuable.

For the figures I thank Linda Bethel and Chris Stanton; Linda and Tina Bernard gave many helpful tips on LaTeX and on various aspects of file organization.

Last but far from least, I offer thanks to two unusual organizations, the Graduate School of Business at Stanford University and the Center for Advanced Study in the Behavioral Sciences, for being such stimulating work environments.

## CHAPTER I

# Introduction

JONATHAN BENDOR

There are two main orientations toward bounded rationality (BR) in political science. The first orientation sees the glass as half full, emphasizing that decision makers often manage to do "reasonably well"—even in complex tasks—despite their cognitive limitations. Virtually all of Simon's work and also the theory of "muddling through" (Lindblom 1959; Braybrooke and Lindblom 1963) belong to this branch, which we can call the problem-solving approach. In the second orientation the glass is half empty: the emphasis is on how people make mistakes even in simple tasks. Most of the research on heuristics and biases, following Tversky and Kahneman's pioneering work (1974), belongs here.[1]

Prominent early use of the problem-solving approach can be found in Aaron Wildavsky's studies of budgeting. In, for example, *The Politics of the Budgetary Process*, he devotes much space to showing how and why making resource allocation decisions in the federal government is beset by complexities *and* how the professionals cope with their difficult tasks: "It [is] necessary to develop mechanisms, however imperfect, for helping men make decisions that are in some sense meaningful in a complicated world" (1964, p. 11). One might argue that his orientation was due simply to the time paths of these different intellectual currents: Simon and Lindblom had launched the problem-solving branch before Aaron wrote his pioneering book on budgeting, whereas the Tversky-Kahneman branch didn't get started until nearly a decade later. But there is a deeper explanation. Aaron did field research on federal budgeting, including 160 interviews with "agency heads, budget officers, Budget Bureau staff, appropriations committee staff, and Congressmen" (1964, p. v). He was not interested in how experimental subjects committed errors of judgment or choice in laboratory settings; he was interested in how real decision makers tackled problems of great complexity. Hence, he was intrigued by how they managed to do this extremely difficult task reasonably well.[2] One sees in the book a respect for the decision makers, arising in large measure from an appreciation of the difficulty of the tasks they confronted.

Indeed, I suspect that the seriousness with which Aaron thought about the tasks of budgetary officials was part of a long-standing theme of his professional life: a passionate interest in the real-world problems confronting government officials in a modern society. (Helping to found Berkeley's Graduate School of Public Policy was another reflection of this theme.)

This is more than biographical detail. It also illustrates an important—though neglected—part of the problem-solving approach to bounded rationality: a close examination of decision makers' tasks. In Simon's pioneering formulation, the focus was always on a *comparison* between a decision maker's mental abilities and the complexity of the problem he or she faces: for example, "the capacity of the human mind for formulating and solving complex problems is very small compared with the size of the problems whose solution is required for objectively rational behavior in the real world—or even for a reasonable approximations to such objective rationality" (1957, p. 198). Thus, for Simon, as for Wildavsky, the idea of bounded rationality is *not* a claim about the brilliance or stupidity of human beings, independent of their task environments. Many social scientists miss this central point and reify the idea of BR into an assertion about the *absolute* capacities of human beings.[3] The fundamental notion here is that of cognitive limits; and, as for any constraint, if cognitive constraints do not bind in a given choice situation, then they will not affect the outcome. *And whether they bind depends vitally on the information-processing demands placed on the decision makers by the problem at hand.* More vividly, Simon has called the joint effects of "the structure of task environments and the computational capacities of the actor … a scissors [with] two blades" (1990, p. 7): Theories of BR have cutting power—especially when compared to theories of (fully) rational choice—only when both blades operate. Thus, any analysis that purports to fall into this branch of the research program yet examines only the agent's properties is badly incomplete.

Thus, Wildavsky not only belonged squarely in the problem-solving branch of the BR program; his intellectual propensities—his interest in how real officials tackle real problems of great complexity—predisposed him to use *both* blades of Simon's scissors. That was unusual. It was also productive: many of his insights about budgeting flowed from his effective use of Simon's scissors.

Of course, every research method focuses our attention on some scholarly questions in the domain at hand *and* deemphasizes others in that same domain. (Lindblom's warnings [1959] about the utopian folly of trying to be comprehensive apply to academics as well as to government officials.) So it is not surprising that Wildavsky's research methods led him to ignore certain topics. In particular, his interest in *applying* the basic ideas of bounded rationality to the study of real-world budgeting steered him away from analyzing the foundations of BR theory. That simply was not part of his intellectual agenda. But a serious focus on those foundations is long overdue. Brilliant as they were, neither Simon nor Lindblom said it all. We political scientists—particularly those of us who work on the behavioral (bounded rationality) side—have done too much quoting and too little reworking. I believe that we will see vigorous scientific competition between rational choice (RC) theories of policy making and behavioral theories only if behavioralists take the foundations of their theories as seriously as RC theorists take theirs. Further, I think that this entails transforming verbal theories into mathematical models. (For an argument on this point in the context of incrementalism, see chapter 4.)

The next section surveys a family of theories that has been central to the problem-solving branch of the BR program: those that use the idea of *aspiration levels* as a major concept.

## THEORIES OF ASPIRATION-BASED PROBLEM REPRESENTATION AND CHOICE

The main claim I offer in this section is that the idea of aspiration-based choice constitutes a major family of theories in the bounded rationality research program. The word *family* matters: I think it is a serious mistake to view satisficing per se as an alternative to theories of optimization. As careful scholars working in the optimization tradition have often pointed out, there is no single RC theory of (e.g.) electoral competition (see, e.g., Roemer's comparison [2001] of Downsian theory to Wittman's), much less just one RC theory of politics. Similarly, satisficing is a theory of search. It is *not* the Behavioral Theory of Everything. Moreover, a key part of satisficing—the idea of aspiration levels—is shared by several other important behavioral theories: theories of reinforcement learning (Bush and Mosteller 1955) and prospect theory (Kahneman and Tversky 1979). So I first argue that a "family" of theories is a significant grouping that fits into the more conventional hierarchy of research program, theories, and models, and that substantively we can gain some insight by focusing our attention on this common feature of aspirations.

(A reason that it is methodologically important to identify this family of theories is that the size of this set is probably indefinite. That is, an indefinitely long list of choice problems may be representable via aspiration levels. I see no reason why this should not hold. All that is required of the choice problem is that there be more than two feasible payoffs, but the agent simplifies the problem by reducing that complex set into two simple equivalence classes. Importantly, there is *no* restriction on the substantive type of problem, or the task of the decision maker, which can be represented this way.)[4]

Technically, an aspiration level is a threshold in an agent's set of feasible payoffs. This threshold partitions all possible payoffs into two disjoint sets: those below the threshold and those that are greater than or equal to the threshold.[5] In all aspiration-based theories, this dichotomous coding *matters*: that is, important further implications flow from this representation of the choice problem.[6] The details of these implications vary, for they naturally depend on the substantive nature of the theory at hand, as we will see shortly. This parallels how the details of optimal strategies vary, depending on the type of problem confronting the decision maker. Strategies of optimal candidate location in a policy space look quite different from strategies of optimal nuclear deterrence. But they share a common core, that of optimal choice. Similarly, psychological theories of learning look quite different from prospect theory, but they too have a common core. Interestingly, we will see that early on, scholars working on certain members of this family of theories were not even aware that their particular theories *required*, as a necessary part of their conceptual apparatus, the idea of aspirations; they backed into this idea. Let me now briefly describe several important members of this family of theories: satisficing and search, reinforcement learning, and prospect theory.

### Theories of Aspiration-Based Behavior

*Search*    Search was Simon's original context for satisficing. The idea is simple. Simon posited that when an agent looked for, say, a new job or house, he or she had an *aspiration level* that partitioned candidates into satisfactory options and unsatisfactory ones.[7] As soon as the decision maker encountered a satisfactory one, search ended. The verbal theory suggests that the aspiration level adjusts to experience, falling in bad times (when one searches without success) and rising in good times (swiftly encountering something that far exceeds the aspiration level), but this was not represented by the formal model. (Cyert and March [1963] allow aspirations to adjust to experience in their computational model.)

There is, however, a rival formulation: optimal search theory. The basic idea is that agents assess both the expected marginal gains and the expected marginal costs from searching further and set an optimal stopping rule that equates the two. Some behavioralists (e.g., Schwartz et al. 2002) are completely unaware that this rival exists. This is a pity. The ignorance allows such behavioralists to have scholarly aspirations that are too low: they think satisficing easily beats RC theories because the latter predict that decision makers search all options exhaustively before making a choice. Since this is obviously false in many (most?) domains, satisficing (hence BR, etc.) comes out on top in this scientific competition. But this horse race was bogus: optimal search theory does *not*, in general, predict exhaustive search. Indeed, in many environments the optimal stopping rule takes the form of a cutoff: if the agent stumbles on an option worth at least $\bar{v}$, stop searching; otherwise, continue.[8] Significantly, the main qualitative features of this prediction are exactly the same as those of the satisficing theory: the set of feasible options is partitioned into two subsets, and the searcher stops looking upon finding something belonging to the better subset. Clearly, then, some additional cleverness is required in order to derive different predictions from RC and BR theories of search. It's doable, but we can't stop with the obvious predictions.

Now that the notion of optimal search has been spelled out, it is clear that in most search problems the stopping rule could be *either* suboptimally low *or* suboptimally high. If the costs of search are sufficiently low, then one should keep searching until one has found the highest-quality option, but this will rarely be the case in policy making, particularly not when decision makers are busy (Behn and Vaupel 1982).

Interestingly, however, the possibility that uncalculated aspirations may be *too high* has gone almost unnoticed by the behavioral literature. Indeed, an auxiliary premise has been smuggled into the concept of satisficing: it is virtually *defined* as search with a suboptimally low threshold (as in the phrase that probably most of us have heard, "*merely* satisfice").

The problem is not with the definition per se—one can stipulate a technical term as one pleases—but with its uses. This implicit smuggling of an important property into the heart of satisficing helps us to overlook the *possibility* of excessively high aspirations. It reinforces the mistake of equating "optimal" with "best quality," or worse, assuming that optimal equals perfect. (Chapters 3 and 5 analyze the implications of equating optimal with perfect.) We thereby neglect some of the empirical content of aspiration-based models of adaptation.

*Learning*    In experiments on learning, psychologists often give subjects a set of options that they can repeatedly try. In so-called bandit problems, every option either pays off some fixed amount $v > 0$ or yields nothing. Some options are better than others—pay $v$ with a higher probability—but subjects aren't told which. Instead, they must learn which options are better.

Originally, these experiments were part of the behaviorist research program in psychology, which eschewed mentalistic concepts such as aspirations.[9] However, learning theorists in psychology discovered (the hard way) that they needed this concept to explain the behavior of subjects.[10]

This issue becomes more pressing in choice situations where there are more than two payoffs. Given only two payoffs, it is quite natural to hypothesize that subjects will regard getting something as a success while getting nothing is a failure.[11] But many choice situations do not provide such an obvious coding. In, for example, the two-person prisoner's dilemma, is the payoff to mutual cooperation reinforcing? How about the payoff to mutual *defection*?

*Prospect Theory*    Perhaps the best-known postulate of prospect theory (Kahneman and Tversky 1979) is that decision makers are risk averse regarding gains but risk seeking regarding losses. This is not, however, a good way to remember the theory. Its fundamental axiom—its most important departure from classical utility theory—is that people evaluate outcomes relative to a reference point.[12] Indeed, under the classical theory, the claim that people are, for example, risk averse about gains *makes no sense*: the idea of a "gain" has no place in the conceptual apparatus in standard utility theory.[13] Decision makers simply have preferences over baskets of consequences; that's all that matters. They do not compare baskets to an internal standard of goodness or acceptability.

A reference point is, in effect, an aspiration level.[14] Of course, aspirations in prospect theory have a different function than they do in satisficing-and-search theory. (Prospect theory has not been applied to search problems, as far as I know.) Rather than serving as a stopping rule, aspirations in the context divide the set of feasible outcomes into those coded as gains and those coded as losses. But again we see a dichotomizing of payoffs into two qualitatively different subsets.

## An Important Problem: The Empirical Content of Aspiration-Based Theories

Although I think that aspiration-based theories form a tremendously important family in the BR research program, no set of theories in the social sciences is free of problems. (At least, I have not been lucky enough to encounter such a set!) And since I completely agree with Martin Landau's view that scientific progress depends tightly on criticism, I think it is vital for scholars who work on these theories to detect their weaknesses and work on them. I do not think that aspiration-based theories of choice are perfect. That would be a reflexively bizarre claim.

Here I want to identify just one problem—but, I think, a significant one. It concerns the empirical content of several kinds of aspiration theories: at a minimum, those of search and of learning. (Here I am relying entirely on the results of Bendor, Diermeier, and Ting [2003, 2007], hereafter BDT.) The problem is simple: many of these theories have very little empirical content. Indeed, some probably cannot be falsified. This is troubling.

These are strong claims; they should be demonstrated. I will sketch out one of BDT's results (which are all deductively established) to give a sense of the logic. Consider an agent who learns via classical trial and error: try an action; if it "works" (the current payoff, $\pi_t$, is at least as big as the current aspiration, $a_t$), then the agent becomes more disposed to use it again. If it fails ($\pi_t < a_t$), then the agent becomes less likely to try it in the future. In accord with the conventional wisdom, aspirations also change with experience, moving up in good times and down in bad. (These informal ideas can be made mathematically precise: see Bendor, Diermeier, and Ting (2003, p. 264) for a formal specification of these axioms of propensity adjustment and aspiration adjustment. Theorem 1, below, refers to these as axioms 1 and 3, respectively.) Then we get the following result.

THEOREM 1. *Consider any repeated game with deterministic and stationary payoffs in which players adjust their action propensities by any arbitrary mix of adaptive rules that satisfy axiom 1 and adjust their aspirations by any arbitrary mix of rules that satisfy axiom 3. Then any outcome of the stage game can be sustained as a stable outcome by some self-replicating equilibrium. (BDT 2003, 2007)*

This is a "folk theorem" for reinforcement learning, in the same sense that there are folk theorems for repeated games in noncooperative game theory: if anything is stable, then the theory isn't predicting much.

Fortunately, there are ways of ameliorating this problem and restoring empirical content to models of aspiration-based search and learning. BDT identify two options: one can either assume that players randomly experiment (this is precluded by classical satisficing) or let them obtain *vicarious* experience by making their aspirations depend partly on other people's experience (payoffs). The latter method builds the venerable sociological idea of reference groups into these models. (Classical satisficing theory is asocial.)

## CONCLUSION: SOME POLICY IMPLICATIONS

The contest between RC and BR theories will last for decades. But life and its problems go on. Given this difference in tempos between basic and applied science, it's important to put some pressure on basic scientists. After all, if the change is as big a deal as the proponents of the challenging research program claim, it should generate interesting policy implications. Below are two that I think are significant. The first implication concerns the normative analysis of public policies; the second is descriptive as well as prescriptive.

### Happiness, Aspirations, and Policy Evaluation

Since Bentham, evaluations of public policies have had a strongly utilitarian cast. But modern applications of utilitarianism, such as cost-benefit analysis, are based not only on the classical theory's normative principles but also on descriptive axioms regarding utility—or, more informally, happiness. As noted earlier, these descriptive axioms do not include the concept of an aspiration level. Scholars working in the new field of hedonic psychology, however, have found that concept very useful indeed in explaining some intriguing empirical findings.[15] (In this sense the largely empirical field of hedonic psychology is linking hands with theories of aspiration-based adaptation discussed throughout this book.) For example, although per capita GNP rose dramatically from 1946 to 1990 in France, Japan, and the United States, "there was no increase in mean reports of [subjective well-being]" (Diener et al. 1999, p. 288). Endogenously rising aspirations explain this Faustian dynamic of doing well but feeling no better.

The implications of these empirical findings and their aspiration-based explanation for policy evaluation could be profound. To cut to the chase: if the subjective well-being of citizens affected by *every* public program is determined by their comparing objective payoffs to their subjective aspiration levels, then cost-benefit analysis is seriously flawed. As the data reported by Diener et al. suggests, it could produce systematic biases. In particular, a better-grounded evaluation of the collective impact of our public (and private) policies might show that though we're getting a lot richer (on average), we're not getting much happier.[16]

More specific normative implications require a specific model of aspiration-based choice. Consider prospect theory with endogenous aspirations.[17] Given the axiom of loss aversion, plus the conventional assumption that aspirations adjust to payoffs with a nontrivial lag, such a model implies that a utilitarian measure of social welfare would be enhanced if public policies cushioned people against the short-term effects of sudden catatrophic losses.

## Expertise in the Policy Process: Discovering and Teaching Effective Heuristics

Some scholars in the problem solving branch of the BR program (e.g., Ericsson and Lehmann 1996; Simon 1999b) have studied expertise closely. A major finding in this area is that most experts labor under the same general mental constraints—such as the limited number of pieces of information that can be held in working memory—as do the rest of us. Thus, they do not overcome bounded rationality; instead, they *finesse* it: they've learned domain-specific methods that are procedurally rational (Simon 1990, 1996). One of these methods is heuristic search.

This assertion is, I believe, inconsistent with the views of many political scientists, who seem to believe in the proposition that *amateurs satisfice but experts optimize*. This belief is based on a misunderstanding of how satisficing in particular and heuristics in general fit into Simon's theory of problem solving. Simon realized long ago that chess is so difficult—the decision tree explodes after only a few moves—that even grand masters must search heuristically (Simon and Schaeffer 1992). Part of the difference—as Lindblom guessed—is that they use powerful heuristics; we duffers do not.[18]

Thus, if we are to teach our students to be genuine experts, in whatever policy field they go into, we must take heuristics seriously.[19] This would be a sharp break with the past. With a few exceptions, theoretically oriented political scientists and most policy scientists trained in economics have been oriented toward *strategies*—especially optimal ones—rather than toward heuristics. The two are quite different. Whereas a strategy (in its technical, game-theoretic sense) is a complete plan of action, heuristics can be incomplete. Heuristics—"get your middle pieces out early," "recipro-cate" (in the iterated prisoner's dilemma), "try a proof by contradiction"—may be valuable pieces of advice even if they aren't complete solutions. Moreover, whereas we are accustomed to making very sharp evaluations of strategies—they are either optimal (or part of Nash equilibria, in strategic settings) or not—heuristics come in many shades of gray. Indeed, we use heuristics when we don't have the slightest idea of what *is* an optimal plan for the task at hand. Learning which are "pretty good" heuristics in a given domain and which are mediocre is an important part of becoming an expert.[20] I think we do our professional-degree students a profound disservice if we pretend that all they will need in the real world are strate-gies. Most of the time they will need heuristics; full-blown strategies will be beyond their (or anyone's) reach. Lindblom's cautions have lost none of their punch.

## CHAPTER 2

# Herbert A. Simon

*Political Scientist*

JONATHAN BENDOR

## INTRODUCTION

This chapter's title should end with a question mark. For understandable reasons of professional pride, we political scientists would like to believe that, taken as a statement, this title accurately describes Herbert Simon's career. But it doesn't. For the last forty-plus years of his amazingly productive life he was a cognitive scientist. All the evidence, both objective and subjective, points to this conclusion. The vast bulk of his publications from 1960 on were on topics in cognitive science, as even a quick check of his curriculum vitae shows. (A complete bibliography can be found at www.psy.cmu.edu/psy/faculty/hsimon/hsimon.html.) And it is not merely the quantitative evidence that reveals the pattern: his original research from 1960 on was clearly in cognitive science. (Indeed, he continued to work on original cognitive science projects in the hospital just before he died [Janet Hilf, personal communication]). In contrast, his political science papers during this long period were usually responses to external stimuli, such as winning the Madison or Gaus awards from the American Political Science Association. Although the Madison and Gaus lectures provide useful overviews of the implications of his research for political science, they do not report ongoing research. As for his self-identity, in his autobiography he stated flatly that by 1956 he had been transformed "professionally into a cognitive psychologist and computer scientist, almost abandoning my earlier professional identity" (Simon 1991, p. 189). The combination of this objective and subjective evidence is compelling. Much as we might like to believe the lovely title of Dubnick's salute (2002) to Simon—"Once a Political Scientist, Always..."—after 1960 he wasn't.[1] And since Simon was famous for his ferocious dedication to the truth—he was no epistemological relativist, and one can easily hear him saying, "If this be positivism, let us make the most of it!"—he would be upset if we deluded ourselves. Facts are facts.

Yet although Simon was not a political scientist for most of his career, his work remains relevant to our discipline. Indeed, it offers far more than has thus far been used. This chapter aims to convince readers that this claim is true—and that it applies to his post-1960 work as well as his earlier and more familiar work. The best is yet to come.

Because most of this chapter is an attempt to convince readers of this, the following should suffice for now. It is clear that rational choice (RC) theories collectively constitute the most important research program in political science today. I believe that this is a salutary development for many reasons, not least of which is the theoretical coherence RC has brought to the field. But there is a major problem here, and by now we all know what it is: decision makers are not fully rational. And positing that they are yields predictions that are, in at least some important political situations, wildly inaccurate; these predictions do not correspond to the facts as we know them. (The literature on violations of expected utility theory is voluminous; see, e.g., Camerer [1994]; for violations of game-theoretic (Nash) predictions, see Camerer [2003].)

There are, of course, methodological replies to this problem. A strategy that is often used in formal theory seminars is a dogmatic and simplified version of Friedman's instrumentalism (1953): one should challenge a model's implications but never its assumptions. But I suspect that even a committed advocate of Friedman's position knows, deep down, that it is not a Good Thing if the premises of one's model are way off the mark. Such a model may have other redeeming features, but then the reasonable evaluation is that the model may he usable, for certain purposes—but *despite* its postulational inaccuracy, not *because* of it.

However, one cannot beat something with nothing: it takes another research program—not just disconfirming evidence—to challenge a well-entrenched incumbent program. Enter Simon. There are intellectual resources in the body of work collectively labeled "bounded rationality" that could create an alternative to the RC program. Achieving this goal will take considerable effort, as such a program is not fully developed in Simon's own work. (It would be remarkable if it were.) Part of this chapter will develop this point: too little of the potential in Simon's work has been tapped. We have often waved a magic wand, labeled either "bounded rationality" or "satisficing," but such wands by themselves do little scientific work. It is, however, possible to construct such an alternative research program.

It is also desirable. Hegemony, put more neutrally, is monopoly—the absence of competition. And as more than a few scholars (Lakatos and Feyerabend in philosophy, Merton in sociology, Landau in political science) have argued, competition among both theories and research programs has many benign scientific effects. Their arguments are sufficiently well-known that I need not repeat them here. Let me say only that if most of the discipline embraces RC substantively while simultaneously embracing wing walking (Shepsle 1996) as a methodological rule, it could create a self-fulfilling trap.

The second section of this chapter offers a short overview of the research program. Next comes a section that describes the program's two main branches in our discipline: Simon's and the Tversky-Kahneman branch. The fourth section briefly examines the somewhat puzzling pattern of the Simonian branch's impact on political science. The fifth turns to some applications within political science of Simon's early (pre-1960) phase. The sixth section reviews some untapped potential in his later work. The seventh concludes.

## BOUNDED RATIONALITY: WHAT IT ISN'T AND WHAT IT IS

A common misconception is to confuse the general idea of bounded rationality (BR) with the much more specific idea of satisficing. (Much of this section is taken from Bendor 2001.) This is sometimes done dismissively, as in the following hallway syllogism: (1) BR "boils down to" satisficing; (2) satisficing is simply a search-for-alternatives theory that takes the costs of computation into account; (3) hence, far from being a major rival to the RC program, BR is just a minor tweak on optimal search theory.

This is a major error that produces a serious underestimation of the program's content. Confusing BR with satisficing conflates three types of symbolic formulations: research programs, theories, and specific formal models of those theories. Research programs typically contain many different theories: RC contains Downsian theories of party competition, theories of collective action, and so on. Some of these theories conflict; others are simply about different phenomena. And many of these theories are formalized as models: for example, Downs's ideas (1957) have given birth to scores of formal models. So to conflate BR with, say, Simon's specific *model* of satisficing (1955) is a methodological howler. Because the BR research program focuses on individual decision making and because the postulated mechanisms can appear in many contexts, its empirical domain is vast—it is no less imperialistic than RC—so its set of possible theories is also very large.

To identify the program's central ideas, one should start with Simon's formulation: "The capacity of the human mind for formulating and solving complex problems is very small compared with the size of the problems whose solution is required for objectively rational behavior in the real world—or even for a reasonable approximation to such objective rationality" (1957, p. 198). Crucially, in this statement BR is a *relation* between a decision maker's mental abilities and the complexity of the problem she or he faces. It is *not* a claim about the brilliance or stupidity of human beings, independent of their task environments. Writers commonly miss this central point and often reify the notion of BR into an assertion about the absolute capacities of human beings, as in the summary that students sometimes give: "Simon thought people are dumb." He didn't. He thought that some of our problems are difficult. This wouldn't matter if our mental capacities had no limits. But they do.

Since this combination of cognitively constrained decision makers tackling hard problems is the program's foundation, it is worthwhile to explore its implications via an example. Consider a group of adults of normal intelligence, randomly paired up to play either chess or tic-tac-toe. A pair of players can either play the game they were assigned to or they can stipulate a particular outcome, take their money for participating in the experiment, and leave. The intuitive prediction is that more people assigned to tic-tac-toe would stipulate an outcome (a draw): after all, that game is trivial for normal adults.

But so is chess, in a sense (Rubinstein 1998, p. 130): Zermelo proved long ago that there exists an optimal way to play chess, and if it is played optimally then—as in tic-tac-toe—the same outcome always occurs: either white wins or black does or they draw. Indeed, for classical game theory, which ignores cognitive constraints, chess and tic-tac-toe belong to the same class: they are zero-sum, finite games of perfect information. Hence, they are game-theoretically equivalent.

In the real world, of course, these two games are not equivalent. Normal adults do not play tic-tac-toe; they find the game pointless. But they do play chess. Indeed, some take it up as a profession, spending thousands of hours on it.

In our thought experiment people were randomly assigned to the two games, so their different reactions could not be imputed to differences in their cognitive capacities. Instead, their reactions had to be due to the huge differences in the complexity of the games. The point is simple but essential: the mental abilities of normal adults are a binding constraint in chess but not in tic-tac-toe. Accordingly, BR and RC theories make observationally equivalent predictions about the latter but not the former. Hence, BR theories have cutting power in chess—knowing the players' cognitive limits gives us predictive leverage—but not in tic-tac-toe.

The example provides a more subtle point. Chess becomes simpler in the endgame, but the players' basic cognitive capacities remain the same. Since the task's demands are falling while mental resources stay fixed, at some point those resources may no longer be a binding constraint. If that happens, the players will play optimally, so a standard RC theory will accurately predict their behavior from that point onward. And indeed this is so: a chess master suffering from a significant material disadvantage in the endgame will resign, because he knows that his opponent will win and nothing can be done about it. Once the players understand what the optimal strategies are, play becomes completely predictable; hence, continuing the game is as pointless as playing tic-tac-toe.

This example reveals the subtlety of the contest between BR and RC theories. Models from these two research programs make different predictions about behavior in chess only when the game is sufficiently complex to make players' mental capacities binding constraints. For expert players, this condition holds early in the game but not at the end. Thus, the two types of theories make observationally equivalent predictions about the *endgame* behavior of experts. Further, novices, not realizing that having one more pawn than their opponent at the end can be decisive, will often play it out to the bitter end. Thus, a BR theory would predict that the crossing point—when cognitive capacities cease to bind—occurs much later, if at all, for novices than for experts. Accordingly, the predictions of the two types of theories will differ more for novices than for experts.

The idea that BR's significance turns on the *difference* between cognitive resources and task demands, not on cognition's absolute level, implies what might be called a scaling principle of modeling: what matters in a model is not so much how sophisticated the agents are assumed to be or how hard the problems are but rather the *difference* between the two. Typically, real humans are more sophisticated than agents in BR models, but real problems are also harder; both are scaled down in models. And as long as a model scales both sides symmetrically, it may be plausible even though the agents in the model are quite dumb.

## Essential Properties of Humans as Information Processors

What, then, are the essential properties of humans as information processors? That is, what properties should be on any political scientist's short list of cognitive features to consider when constructing behaviorally plausible theories of political decision making?[2] (For a much more thorough examination of these and related properties of human information processing, see Jones 2001, pp. 3–127.)

Synthesizing Hogarth (1987), Simon (1990), and Lodge (1995) gives us the following list of attributes:

(1)  Perception of information is selective: there is always far more information in our objective environments than we can perceive or attend to. Thus, perceptions must be strongly guided by anticipations. The main point here is not necessarily bias but *top-down processing*: perception is influenced by schemas and other mental constructs.

(2)  High-order information processing, especially conscious thinking and attention, is largely serial in nature.[3] This has significant implications for the real-time behavior of busy operators, managers, and executives (to use James Q. Wilson's trichotomy [1989] of public officials).

(3)  Compared to modern computers, humans process many kinds of information slowly. This is partly due to physiological limits: neurons transmit impulses much more slowly than do electrical circuits.

(4)  People are not strong at calculation, compared to computers or even calculators.

(5)  Memory is not photographic; it is actively reconstructive (Schachter 2002). It therefore lacks photographic fidelity.

(6)  Although there is no known limit to long-term memory, short-term or working memory is very small (Miller 1956; Cowan 2001). Thus, because everything that enters the former goes through the latter (Miller 1989, p. 151), the working memory is a key bottleneck in information-processing.

## Theoretical Implications for Political Science: A Caveat

It is important for political scientists—properly concerned with much more aggregate phenomena than cognitive psychologists are—to understand the theoretical implications of the above properties for social scientists. For a political scientist interested in modeling a large-scale phenomenon such as voting, it is not necessary that the model explicitly represent the above constraints.

Instead, what is fundamental is that the postulated decisional capacities of agents be consistent with these limits. For example, something is wrong if a model assumes that decision makers have perfect recall of the complete history of a long and complicated game (see constraint 5 above); something is *seriously* wrong if the model's implications depend sensitively (i.e., in a knife-edge way) on such an assumption. In contrast, if a specialist on public agencies wanted to model the behavior of a single official—an air traffic controller or a general—then it seems appropriate to model these mental limits explicitly.

### BR'S MAIN BRANCHES IN POLITICAL SCIENCE

As noted in chapter 1, Simon's version of bounded rationality focuses on how we cope with complex problems despite our cognitive constraints. In contrast, the heuristics-and-biases approach, pioneered by Kahneman and Tversky, has emphasized how easily we can err even when our tasks are simple. Although the distinction is partly one of framing—the two branches agree about much and the tradition has acknowledged its ties to the Simonian program—it is nonetheless significant and warrants elaboration.[4] (For a cogent analysis of the intellectual antecedents of the T-K branch, especially regarding its focus on human error, see Hammond 1990; For valuable overviews, see Dawes 1998 and Gilovich and Griffin 2002.)

## The Glass Is Half Full

As noted above, one of Simon's key premises is that a decision maker's "inner environment"—his or her cognitive constraints—will "show through" only when the decision problem is sufficiently difficult.[5] When these constraints do not bind, agents will optimize, adapting completely to the demands of the outer environment and revealing only their preferences in the process.[6] This idea naturally suggests that one should study people facing hard problems, for it is in such circumstances that one will empirically discover the important types of mental limits, such as that on short-term memory (Simon and Chase 1973). Thus, it is no accident that chess has been a fertile topic for this branch of BR: its combinatorial explosion ensures that no one (and no thing) can play chess optimally; information-processing constraints must show through.

But studying chess revealed more than mental limits. An interesting (though possibly unintended) consequence of examining this game and similar hard problems has been the generation of new questions for the Simonian line, especially about performance. In these domains people's performance varies tremendously: a few are grand masters, many are duffers. This variation cries out for explanation. Thus, a new research focus emerges: how do some decision makers do so well, given that they labor under similar cognitive constraints (Simon and Simon 1962)? Hence, this branch came to see that the glass is at least half full: in some undeniably hard domains, some human beings do indeed perform "reasonably well" in terms of an empirically sensible aspiration level.[7] (The theoretically defined aspiration level of optimal play is irrelevant for an empirical theory of chess, since everyone, from Kasparov and Deep Blue on down, flunks this test. Hence, using this criterion produces no variation in the dependent variable of performance.)

What has happened, then, is that researchers in the Simonian branch have in a sense come to take BR for granted—not in thinking it unimportant but in believing that it needn't be established. There is simply no doubt that mental constraints bind in the problems subjects work on in this branch: the exploding decision trees preclude objective optimization. What is interesting is not whether the behavior of subjects is fully rational— we know it isn't—but how relatively competent agents work around their (indisputable) cognitive limitations.[8]

In public administration, certain jobs—for example, those of air traffic controllers on busy days or generals in the fog of battle—are clearly hard. Cognitive constraints should show through in such roles, yet human performance varies substantially. Thus, the general lessons of the study of chess (not, of course, the detailed conclusions about domain-specific heuristics) are applicable to such roles studied by our discipline.

## The Glass Is Half Empty

Ward Edwards, a pioneer of behavioral decision theory, once remarked that "A research focus on systematic errors and inferential biases can lead those who read the research with an uncritical eye to the notion that such errors and biases characterize all human decision making" (von Winterfeldt and Edwards 1986, p. 530). Although Edwards went on to say that "a sense of the inevitability of specific kinds of intellectual errors has been more widely disseminated than, we believe, the literature proposes or justifies" (p. 531), his remark still accurately describes how many social scientists *perceive* the heuristics and biases branch of BR (Christensen-Szalanski and Beach 1984).[9] This was probably not Tversky and Kahneman's intention. Their idea was to map the subtle processes of mental framing that can cause cognitive illusions, similar to the study of perceptual illusions: "In the persistence of their appeal, framing effects resemble visual illusion more than computational errors" (Tversky and Kahneman 1986, p. 260). This project's results are more striking if they can be shown to occur even in simple tasks, just as perceptual illusions are more striking if they can be demonstrated with simple stimuli; it is more striking still if one shows that even experts are vulnerable to these cognitive illusions. Thus, this branch of BR went down a markedly different path: in prototypical investigations, experimenters showed that even highly trained subjects answering simple questions perform suboptimally—a sharp contrast to the Simonian program's emphasis on "good" performance in difficult domains.

These differences indicate that in one respect the T-K branch has set for itself a more ambitious research agenda. Obviously, it should be harder to demonstrate that our cognitive limits show through on simple problems than on hard ones. Thus, the T-K program has worked on pushing out the boundaries of BR by showing that even quite subtle problem representations ("framing") can induce suboptimal performance.[10] One needn't go all the way to chess to uncover our mental limitations: it turns out that humans are sufficiently sensitive (hence vulnerable) to framing that judgmental or decisional imperfections will appear even when a task exhibits no combinatorial explosion at all. Thus, one of the main findings of the T-K branch is that the empirical domain of the BR program—where it has predictive bite—is larger than once believed.

Naturally, if one is pushing the boundaries of an idea or hypothesis, one is more likely to step over the line, which in this case means overestimating the boundaries of BR. Unsurprisingly, there has been a scholarly backlash, attempting to show that humans are better decision makers than the T-K program makes us out to be (e.g., Gigerenzer 1991; for a reply, see Kahneman and Tversky 1996). Unfortunately, this debate has been clouded by a misunderstanding of Tversky and Kahneman's original intentions, which was not to show that *Homo sapiens* are dolts but rather to uncover fundamental cognitive mechanisms that leave their imprint nearly everywhere.[11]

It must be acknowledged, however, that some of the responsibility for this clouding falls on some of the scholars who work in this tradition: their "gotcha" approach to their subjects and to experimental design gives the impression that the goal is to show that humans are, in fact, inept decision makers. (For a critique along these lines, see Lopes 1981.)

This tendency has been exacerbated by the fact that specialists working in the T-K branch have shown little interest in performance variation. Little in this line of work has been equivalent to the study of chess masters.[12] As noted, this made sense, given the primary objective of demonstrating the nearly ubiquitous character of certain mental processes. But it did have the unfortunate side effect of making the program largely—not completely—indifferent to natural variations in performance.[13] Instead, the evaluations are usually based on dichotomous theoretical standards: do people make choices in accord with the axioms of expected utility theory? do they revise beliefs in a Bayesian manner? The answers—generally no, they don't—are less informative than would be answers based on a quantitative scale that measures degrees of sophistication, even if the best empirical performance falls far short of the theoretical ideal.

Finally, the intellectual differences between the two branches are reflected in and reinforced by differences in citations, publication patterns, and other aspects of scholarly practice. Work in the Simonian tradition on problem solving in difficult tasks rarely cites T-K studies. Conversely, beyond brief mentions of Simon's classical articles on bounded rationality of the mid-1950s, T-K studies of human error in simple tasks rarely cite work in the Simonian tradition.[14] It is strange: Simon, known for his pioneering work on behavioral theories of choice, is now rarely cited in surveys on behavioral decision theory. (For two exceptions to this pattern, see Abelson and Levi 1985 and Slovic et al. 1988. Both devote a nontrivial amount of space to the Simonian program.) By "decision making," the literature apparently means behavior in *simple* choice environments. If the task is complex, it is given over to the field of problem solving. Yet this division is artificial: in reality, problem solving and choice are closely connected. Treating them as if they were does not "cut nature at the joints."

THE SIMONIAN BRANCH'S IMPACT ON POLITICAL SCIENCE

The Simonian branch has had a peculiar pattern of impacts. Four facts stand out.

(1)  By now Simon's thinking has had a bigger impact on theorizing in economics than in Simon's own field of political science.

(2)  Even this impact took decades, however.

(3)  Political scientists, now preoccupied with developing the RC pro-
     gram, which originated in economics, are largely unaware of how
     intensely high-brow economic and game theorists are working on
     BR models.

(4)  Simon's biggest impact was on a field that did not exist before he
     started: the cognitive sciences.

I elaborate on the first three points below.

The Simonian program has had a relatively modest impact on political science. True, Simon himself has been extensively cited by political
scientists, especially by specialists in public administration. Moreover, his
reputation in the discipline is very high. Yet if we inspect the literature in
political science for work that either tests his ideas empirically or develops
them theoretically, the pickings are lean.

The clearest indication of the lack of theoretical development can be
found in the discipline's approach to Simon's most famous idea, satisficing.
As set out in his famous 1955 paper, the early theory of satisficing was not
applicable to many problems in politics. The reason was simple: whereas the
theory analyzed a single, isolated decision maker, political science focuses
on multiperson situations. There was nothing wrong with the initial formulation being decision theoretic; indeed, one could make a good case, based
on an incremental strategy of scientific progress, that it was exactly the
right first step in developing a behaviorally realistic theory of choice. The
problem was the discipline's reaction. Instead of treating the theory and its
formalizations (1955, 1956) as work in progress, the first in what should
have been a long series, the discipline largely treated the theory as a finished
product. Hence, few political scientists in the following decades constructed
theories of satisficing that were more appropriate to political contexts.
Worse, even some thoughtful scholars saw "the Simon-March tradition
[as having] been ... thoroughly nonpolitical in its design and development"
(Moe 1991, p. 111).[15] Given that many saw the formulation as a finished
product that required no active work, while others saw it as apolitical, it
is perhaps not surprising that the Simonian program stagnated in political
science.

The reaction in economics was different, though it too produced a lag.
Because economics already had a well-established research program, built
on the twin foundations of individual optimization and market equilibrium,
and because Simon's program challenged a core postulate of RC, the initial

reaction of most economists was hostility or, at best, indifference. Cyert and March's significant work, *A Behavioral Theory of the Firm* (1963), was designed to show economists what the new research program could do, but it too failed to generate much interest. There were a few exceptions to this pattern (notably Nelson and Winter 1982), which tended to come from elite economists, but these were just that—exceptions.

Eventually, however, elite opinion had its way. Prominent economists (Arrow 1974) and game theorists (Selten 1989) came to believe that cognitive constraints had to be incorporated in theories of choice. Following them, we see a spate of modeling papers in the last decade, some directly influenced by Simon (the citations are usually to 1955 or 1956), others more indirectly.[16] Today, we see titles such as *Modeling Bounded Rationality* (1998), by a top game theorist, Ariel Rubinstein. So after a long lag, Simon finally had a big impact on this discipline. (For a thoughtful overview, see Conlisk 1996.)

Why is Simon's program, and more broadly BR, finally flourishing in economics but not in political science? A sociology-of-science factor has already been alluded to: persuasion by elite economists. This combines with another sociological property: the status hierarchy in economics is clearer than it is in political science, so the opinions of a few elite economists matter a lot. Finally, there is a crucial difference in human capital between economics and political science: the former had many highly trained formal theorists who, in the 1980s and 1990s, saw how to construct mathematical models of bounded rationality—and how to use these in models of strategic interaction. (For a good bibliography, see Rubinstein 1998.) Simon himself was not too keen on thus line of work (Rubinstein 1998, pp. 188–90), regarding it as insufficiently grounded in our empirical understanding of cognitive constraints, but it is nonetheless clear that this ability to develop mathematical models of BR has made working on the program much more attractive to economic theorists than it otherwise would have been.

This burst of activity has gone largely unnoticed by political scientists. This is ironic. While Simon, trained as a political scientist, is finally having an impact in economics, theoretically oriented political scientists are borrowing RC ideas from economics!

## SOME APPLICATIONS OF THE EARLY SIMON

A research program can make serious scientific headway only by demonstrating that it has problem-solving power (Laudan 1977). There are two kinds of such demonstrations. The first is a hurdle: a new program must show that it can handle already-solved problems—for example, account for an empirical regularity already explained by other programs. The second kind involves novelty: the program shows that it can handle an unsolved problem. Ideally (Lakatos 1970), a theory of a new research program solves all of the problems handled by the incumbent program plus some new ones. The challenging program would then dominate the incumbent in terms of problem-solving power. Not surprisingly, such clear victories are rare (Laudan, Laudans and Donovan 1988). The following topics are organized roughly chronologically, by the date of the first contribution on the topic.

## The Politics of Budgeting

For a long time the study of budgeting was an intellectual backwater in political science. Then Aaron Wildavsky went to graduate school at Yale, was influenced by Lindblom's ideas, and the rest— including *The Politics of the Budgetary Process*—is history.

This summary of how it all began is not inaccurate, but it is misleading about the ending. For as Green and Thompson (2001) show, the path of this subfield is a curious one. The line of work on budgetary politics spawned by Lindblom and Simon was, by many scientific criteria, quite successful. Yet recently it has not flourished. This subsection describes how the BR research program bore fruit, but the rest of the discipline has not responded with the enthusiasm that one might have expected. (For a more thorough account of this story, see Green and Thompson 2001.)

The study of budgeting has been tied to the BR program since its inception. As Simon recalled in his autobiography (1991, p. 370), his experience studying budgeting in Milwaukee was a formative one:

> The budget process ... involved the division of funds between playground maintenance, administered by one organization, and playground activity leadership, administered by another. How was this division (which was a frequent subject of dispute) arrived at? My previous study of economics provided me with a ready hypothesis: Divide the funds so that the next dollar spent for maintenance will produce the same return as the next dollar spent for leaders' salaries. I saw no evidence that anyone was viewing the decision in this way. Was I surprised at their ignoring accepted economic theory? Perhaps, initially, but on reflection, I didn't see how it could be done. How were the values of better activity leadership to be weighed against the values of more attractive and better-maintained neighborhood playgrounds?
>
> Now I had a new research problem: How do human beings reason when the conditions for rationality postulated by neoclassical economics are not met?

Although this experience evidently had a major impact on Simon's think-ing about BR in general, it was not applied to budgeting in depth until Wildavsky turned his hand to the topic. (This much of the above bald sum-mary is true.) Since Wildavsky used Lindblom's, "The Science of 'Muddling Through'" (1959), we must first briefly examine that famous paper. Lind-blom and Simon had clearly been thinking on parallel tracks throughout the 1950s. (See Dahl and Lindblom 1953, pp. 82–86; Lindblom, personal communication 2002.) Both believed that the classical model of rationality was often inaccurate as a descriptive theory of choice. But because there are many ways to think about how cognitively constrained decision makers might go about their business—there are many ways of being imperfectly rational—it is not surprising that their positive alternatives differed in sig-nificant ways. In particular, although both maintained that human beings must resort to heuristics to cope with a complex, confusing world, they focused on different kinds of heuristics. (Indeed, it is reasonable to view Lindblom's verbal theory as a set of claims about why perfect rationality is infeasible for complex problems and as a collection of heuristics and a set of claims about their performance. See Lindblom 1968 for a for-mulation that most supports this interpretation. The list on pp. 24–27 is essentially a set of heuristics.) Although Lindblom did not concentrate on one heuristic as much as Simon did on satisficing, perhaps his best-known one was incrementalism—the idea that people would search for solutions in a neighborhood of the current policy.

"Incrementalism" became, of course, the name for Wildavsky's the-ory of budgeting. There were good reasons for this label. Wildavsky himself emphasized the incremental nature of budgeting in the United States, both in the above process sense and in an outcome sense (bud-gets in one year differ only marginally from those of the previous year). And in the quantitative models that Wildavsky developed with Davis and Dempster, incrementalism was given center stage (Davis, Dempster and Wildavsky [DDW] 1966, 1974). But when one inspects the verbal theory of *The Politics of the Budgetary Process*, one sees that the early Wil-davsky was, like Lindblom, more a fox than a hedgehog, to use Isaiah Berlin's terms. True, like any good BR scholar, Wildavsky recognized two big related facts: the complexity of the task environment he was study-ing ("Life is incredibly complicated and there is very little theory that would enable people to predict how programs will turn out if they are at all new" [Wildavsky 1979, p. 9]) and decision makers' ensuing need for "aids to calculation:…mechanisms, however imperfect, for helping [them] make decisions that are in some sense meaningful in a complicated world" (p. 11). But after noting these two important facts, Wildavsky turned with gusto to describing the heuristics themselves (pp. 11–62). And here the fox appeared. No elegant axiomatic theory was given. Instead, he provided, as did Lindblom, a list of techniques that make allocation decisions manageable. (Notably, he included satisficing.) And although he did single out incrementalism, calling it "the major aid for calculating budgets" (p. 13), he covered a broad range of mechanisms. (Interestingly, he included one—role specialization, including the budgetary advocacy of program chiefs—described by Simon long ago in his Milwaukee study.) Thus, at this point Wildavsky's verbal theory was more committed to the premise that decision makers needed *some* aids to calculation than it was to the empirical generalization that incrementalism was usually the heuristic of choice in mid-twentieth-century American budgeting. (In this regard, it is worth noting that he introduced the topic of incrementalism by remarking, "It is against a background of the enormous burden of calculation that the ensuing description...of incrementalism should be understood" (p. 13).) Thus, as of 1964, Wildavsky was a hedgehog about the underlying theory but a fox about heuristics—and most of his book examined the latter.

Later, however, the stance of *The Politics of the Budgetary Process* was obscured. Wildavsky's budgetary theory became *identified* with incrementalism. This was partly due to polemics: much of the critical attention focused on this part of the argument. (The more careful critics did, however, distinguish between process and outcome incrementalism.) But part of the blame falls on Wildavsky himself: when he and his coauthors developed their quantitative model of incrementalism, much of the richness of the verbal theory was lost.[17] This is a necessary price of any formalization, but in this case the price was steep: DDW's quantitative models reinforced the perception that the empirical generalization about local search, or worse, local outcomes, was the theory's central claim. Thus Wildavsky, a brilliant fox, was turned by critics—and by some of his own replies to criticism—into an apparent hedgehog.

Fortunately, other scholars working on budgeting recognized that formal BR theories of budgeting needed to be rescued from quantitative models of incrementalism. Two of the most important efforts were those of J. P. Crecine (1969) and his student John Padgett (1980a, 1981).[18]

Crecine restored Simon's problem—how do cognitively constrained humans deal with the complexities of budgeting?—to the center of the formal theory of public resource allocation. He did so in a novel way, emphasizing the hierarchical nature of budgeting. He meant "hierarchy" not only in the sense of authority but also in Simon's sense (1962) of nearly decomposable systems. The idea was straightforward. An exceptionally common heuristic is to break a complex problem into a set of smaller, simpler ones, hoping that if we solve the smaller ones, we will in effect have solved the original big one. In federal budgeting, this heuristic is realized both by the formal process—different governmental departments are given budgets, as are their subunits, and so on down the hierarchy—and by how decision makers think about resource allocation. Unlike the normative decision rules of public finance that lurked in Simon's mind when he studied Milwaukee budgeting, marginal value comparisons between disparate programs rarely occur in Crecine's descriptive theory. Such comparisons are precluded by hierarchical decomposition. In this theory, decision makers *do* compare spending on guns versus butter, and within defense spending they compare the army's allocation to the navy's. But they do not compare the spending on maintenance for tanks against spending for computers in the Social Security Administration. Such comparisons are never made.

Crecine also argued that the hierarchical decomposition of budget decisions helped top executives control budgetary totals while remaining aloof from the process's details. This point, and the stress on the usefulness of the decomposition heuristic, was especially important in his later work on defense budgeting (e.g., 1970, pp. 231–37) and federal budgeting in general (1985, pp. 114–16, 124–27), in which he argued that for reasons of time and related cognitive constraints it was impossible even for workaholics such as Lyndon Johnson to know much about the budgets of many programs. But they *could* attend to what Crecine called the "Great Identity" (revenues + deficit [or − surplus] = defense expenditures + nondefense expenditures) and try to impose their preferences on the big macroeconomic issues and the guns-versus-butter trade-offs. Similarly, the secretary of defense could try to impose his preferences on big interservice allocation issues, while letting his subordinates worry about intraservice decisions. And so on down the line.

Much of Padgett's model is devoted to how executive-branch officials search for and test alternatives. To understand the significance of his analysis, one should compare it to incrementalist theories of the search for alternatives, in which officials search in the neighborhood of last year's appropriation. This is only one way of cutting the decision tree down to a manageable size. As both Lindblom and Wildavsky recognized, there are other heuristics that can narrow search. BR implies that search is *limited*; it need not always be *local* (incremental).[19] Thus, Padgett's model was an effort to close the gap between verbal theories of budgeting and their formalizations. He assumes a stochastic process of search for budgetary alternatives that is consistent with the general premise of BR but that does not imply that requests are always incremental modifications of the status quo budget. Thus, his model implies a distribution of budgetary outcomes that has "fatter tails" than DDW's, a prediction with empirical support. (For another model of boundedly rational budgeting that yields this implication of fatter tails, see Jones 1999.) Thus, rather than relying (as do DDW) on assumptions about random shocks hitting a deterministic system, Padgett builds the stochastic features into the heart of the model—that is, into the search for budgetary alternatives.

Crecine's and Padgett's achievements were impressive. They rescued Wildavsky from himself by creating formal theories of budgeting that were truer to the informal theory of *The Politics of the Budgetary Process* than Wildavsky's own attempts were. Moreover, their models generated quantitative predictions, and by social science standards did so very successfully. Theory, predictions, corroboration—what more could one ask? So as of the early 1980s, one might have predicted a bright future for behavioral theories of budgeting. Oddly, this did not materialize. Green and Thompson, in their admirable study of this line of work, summarize the current situation as follows: "Mentions of the organizational process model often draw blank stares from students of public budgeting. . . . Bibliographical searches yield only a handful of references or citations. And most contemporary texts entirely overlook the topic"(2001, p. 55).

This is certainly an unusual episode in political science. Rarely in our discipline's history have such successful models been so neglected. (It is rare simply to encounter such successful models, independently of their reception.) While it is beyond the scope of this chapter to explain this trajectory in detail (the interested reader should see Green and Thompson), three reasons for it are worth mentioning because they pertain to some important themes of this chapter.

(1) *Human capital and tools.* Crecine used computer modeling. Very few political scientists were trained in this method. Padgett used analytical stochastic models; even fewer scholars in our discipline understood these techniques.

(2) *Substance and method.* Most political scientists who were mathematically adept were substantively committed to RCT theory. In short, the set of people who belonged to the intersection of the two sets (substantive and skills) was very small. Too small, it turned out.

(3)   *Inaccessibility.* Crecine published in obscure places. (I do not know why.) Padgett's work was inaccessible for the above human-capital reasons. So the two best exemplars of this body of work were either invisible or inaccessible.

*Individual versus Organizational Rationality*

In his first major publication, *Administrative Behavior*, Simon hypothesized—contrary to the American stereotype of bureaucracies as bumbling, mindless systems—that organizations routinely ameliorate the information-processing constraints of individuals. Because the implications of bounded rationality for organization have often been misinterpreted, it is worth quoting Simon so readers can see his position for themselves.

> The argument of the present chapter can be stated very simply. It is impossible for the behavior of a single, isolated individual to reach any high degree of rationality. The number of alternatives he must explore is so great, the information he would need to evaluate them so vast that even an approximation to objective rationality is hard to conceive.
>
>    ... A higher degree of integration and rationality can, however, be achieved, because the environment of choice itself can be chosen and deliberately modified. Partly this is an individual matter ... *To a very large extent, however, it is an organizational matter.* One function that organization performs is to place the organization members in a psychological environment that will adapt their decisions to the organization objectives, and will provide them with the information needed to make these decisions correctly. In the course of this discussion it will begin to appear that *organization permits the individual to approach reasonably near to objective rationality.* (1947, pp. 79–80; emphasis added)

It is a striking fact of intellectual history that Simon's viewpoint was not only ignored, it was *inverted* by several important social scientists. Perhaps the most prominent example of this inversion was in Allison's *Essence of Decision* (1971). The part on the organizational process model summarized the work of Simon, March, and Cyert. Unfortunately, however, this summary completely misunderstood Simon's analysis of how organizational properties compensate for the cognitive constraints of individual decision makers.[20] Instead, it fell back on old, culturally familiar stereotypes about the negative effects of organizational routines in particular and of bureaucracy in general.

Simon's position was, however, taken seriously in Landau's pioneering work on organizational redundancy (1969). Landau showed how structural redundancy, far from being inevitably wasteful, can make an organization more reliable than any of its subunits or members. Consider, for example, an agency working on a difficult R&D project. Suppose the probability that a single team successfully develops a key component is $p$, where $0 < p < 1$. Given the component's centrality and the project's significance, $p$ is considered unacceptably low. But if the agency assigns the problem to $n$ teams working independently in parallel, then the probability that the *organization* will succeed is $1 - [(1 - p)^n]$, which rises steadily toward $1$ as $n$ increases.[21]

The problem is harder when institutions can make two types of error—for example, convicting an innocent person or freeing a guilty one. Here Condorcet's jury theorem comes to the rescue. He showed that *both* types of error could be reduced as juries grew in size, if they operated by majority rule. (For an accessible introduction to the jury theorem, see Grofman and Feld 1988.) Further, his result holds even if the judgmental competency of individual jurors' barely exceeds that attained by pure chance (Grofman, Owen, and Feld 1983; Ladha 1992). Thus, owing to a variety of cognitive constraints, individual decision makers may be extremely fallible, yet the overall organization may be very reliable.

For extensions of Condorcet's result to settings involving conflict (e.g., elections), see studies by Miller (1986) and Grofman and Feld (1988). Their work shows that, in view of the fallibility of individual voters, majority rule can be justified as a collective choice procedure because of its role in reducing error. However, Kinder (1998, pp. 799–800) gives some reasons to restrain one's optimism about this effect of aggregation.[22]

## Satisficing and Search Models

In most of the other applications of the pre-1960 Simon, the notion of satisficing is central. The heart of satisficing models is the assumption that a decision maker has an internal standard, an aspiration level, which partitions all current payoffs into two sets: satisfactory and unsatisfactory. Though this seems like a modest notion, it is actually a major departure from expected utility theory. Let us see why. In standard expected utility theory, a decision maker has a complete preference ordering over all outcomes, which induces a preference ordering over actions. Thus, all that matters is how baskets of consequences compare to each other. There is no sense of the absolute value of any outcome or of the actions that generate outcomes. (It takes students a while to learn this: in this respect expected utility theory is not simply a formalization of ordinary intuitions about decision making.) For this reason, adding or subtracting a positive constant to all the outcomes doesn't fundamentally change the choice situation at all. A theory that posits an aspiration level is a different animal. For example, an agent with unrealistically high aspirations is dissatisfied with anything that happens. Nothing in the classical theory parallels this possibility. The agent's best option is just that—his or her best option—and that's the end of the matter.

This notion has many implications in different contexts. The original context Simon worked on was a one-shot search problem: a person is looking for a solution to a problem and (given satisficing) stops the search upon finding one that meets or exceeds his or her aspirations. But there are many others: this idea can travel to a great many political contexts. In recent years the domain of elections has received special attention.

*Parties in Competitive Elections*    One of the most famous of all RC predictions is that of the median voter theorem in two-party races (Downs 1957). In the standard Downsian setting—in which voters have single-peaked preferences and vote sincerely, candidates are office-seeking, and so on—how would adaptively rational parties behave? (This is a hurdle test. Given the standard setting, the logic of converging to the median voter is compelling; hence, a new theory should pass this test.)

In a primitive early simulation model, Bendor and Moe (1986) showed that if a party used a simple hill-climbing heuristic, then it would converge to the median voter. In this model, however, only a single party was actively adapting, so the electoral landscape was unchanging and the active party's problem was simple. In an important pair of papers, Kollman, Miller, and Page (KMP 1992, 1998) show that both parties can eventually adapt, although only the challenger adjusts in any given election. (Importantly, the winner satisfices, keeping the platform that got him or her elected.) KMP's model ambitiously assumed a multidimensional policy space, but when it is rerun in a standard unidimensional space, the parties converge to the median voter (Page 1999, personal communication).

Since KMP's model is computer run, the researchers had to assume that the challenger used a specific search heuristic (hill-climbing, random search, or an artificial intelligence heuristic called the genetic algorithm). But their result for the unidimensional space generalizes to a large class of heuristics. Bendor, Mookherjee, and Ray (2006) explore this issue analytically. Consider the following model-sketch. Suppose there are $n$ platforms, arranged left to right. As in KMP, incumbents satisfice while challengers search. Apart from the assumption that there are only finitely many platforms, the electoral environment is standard Downsian. In this setting, what are some general assumptions about the challenger's search that still imply convergence to the median? The following spare conditions do most of what we want: they ensure that eventually the *winner's* position, hence governmental policy, converges to the median. So long as in every period there is some chance that the challenger will experiment—that is, take up a platform that he or she has never espoused before—the trajectory of winning platforms must converge to the median voter's ideal point (proposition 3, Bendor, Mookherjee, and Ray 2006).

Apart from this rather weak condition about the possibility of experimentation, one can assume anything at all about search. For example, search could be blind: the challenger could be equally likely to propose any of the feasible platforms. Or, following incrementalism (Lindblom 1959), search could be local, confined to the platforms closest to the challenger's current position, with a hill-climbing heuristic guiding his or her adjustment. And so on. Under these conditions, eventually at least one party will be at the median voter's ideal; the loser may wander around, but it too will find its way back to the median eventually. Thus KMP's unidimensional result can be generalized significantly. And a key part of Downs's median voter theorem can be recovered even if the candidates don't know the location of the pivotal voter or aren't fully rational.

Why does even *blind* search by challengers, plus satisficing by incumbents, suffice? To see why, let us break the process down into two phases: generate and test (Simon 1964). In these models, candidates generate alternatives (platforms) and voters test them. Because citizens always vote for the platform they prefer, the test phase is perfect: for any pair of competing platforms, the winner must be (weakly) preferred by the median voter. Hence, elections cannot hurt the median voter. The worst that can happen in an election is that the challenger fails to find something that the median prefers to the incumbent's position. Hence, all that is required of search by challengers is that *eventually* they find a winning platform. For this, completely blind generation suffices.

The KMP models and their analytical generalizations are a useful illustration of how Simon's early decision-theoretic model of satisficing can be extended to a quintessentially political context yet remain tractable. The modeling trick is to allow the satisficing decision rule to freeze one of the actors—the incumbent—in a natural way: a winning platform ain't broke, so why fix it? Then, although the *context* retains its strategic character (each party's vote share is a function of the other's position as well as its own), the *analysis* in any given period can focus on a single agent's actions and so is quite manageable.

These results tell us that the task environment of two office-seeking candidates in unidimensional policy spaces is sufficiently simple, and the feedback sufficiently clear, that few of the decision makers' cognitive constraints show through in the long run (Simon 1996). The only one that does, because the models allow search to be blind and memory impoverished, concerns the *speed* of adaptation.[23] At any given time, the electoral landscape has "nice" properties: the left party's vote share increases monotonically as it moves toward the right party's position (provided it doesn't overshoot) and decreases monotonically as it moves away. These properties ensure sharp feedback. And even if the challenger ignores this clear feedback, the decisiveness of the median voter ensures that incorrect adaptations cannot take hold. So the overall *system* has enough rationality to ensure that a sensible policy outcome happens eventually.[24]

Similarly, work in economics on markets with "zero-intelligence traders" show that in certain kinds of markets (especially double auction markets), the disciplining effect of markets and market rules suffice to generate aggregate allocative efficiency; individual optimization is unnecessary. Put more quantitatively, "In its first-order magnitude, allocative efficiency seems to be a characteristic of the market structure and the environment; rationality of individual traders accounts for a relatively small fraction (second- or third-order magnitude) of the efficiency" (Gode and Sunder 1993, p. 120). (Given his emphasis on the interaction between a decision maker's task environment and problem-solving capacity, it is not surprising that Simon saw this point. In *Sciences of the Artificial* (1996, pp. 32–33), he explains how the market environment can shape the behavior of even quite "stupid" buyers and sellers, per the Gode-Sunder study.)

A multidimensional policy space rarely has a median voter. This means, under standard game-theoretic assumptions, that there is no Nash equilibrium in pure strategies: given any position by one party, there always exists another platform that could beat it. The lack of a median voter implies a cognitively more difficult task environment (Kotovsky, Hayes, and Simon 1985).[25] Simon's basic argument (1996) implies that more of the agents' adaptive properties should show through in this environment, which is what KMP found (1992, 1998). Heuristics such as hill climbing tend to hang candidates up on local optima, where platforms are better than others in their neighborhood but inferior to a more distant position.

This naturally raises several questions (which collectively form a mini research agenda). First, what factors create more local optima, thus making electoral landscapes more difficult for adaptive agents to handle? KMP identified one property: local optima are most common when voters' intensities on different issue dimensions are statistically independent of the location of their ideal points (1998, pp. 145, 152). This was a genuine discovery: finding it without their model would have been difficult. Second, what internal features of parties make it harder for the party to adapt to the shape of its electoral environment? A clear example of such a constraint is a strong ideological commitment to one's platform; the British Labor Party in the Thatcher era comes to mind. Of course, this is not a purely cognitive constraint in the same way that limits on working memory are for individual decision makers, but it certainly makes it harder for the party to mold itself to the task environment. Further, ideology can become a set of blinders (Lippmann 1922), so it does have a strong cognitive component.

*Boundedly Rational Voters and the "Paradox of Voting"*   The preceding application of the early Simon to competing parties, though important, is almost too easy. It is intuitively clear that in a "nice" (unidimensional) Downsian environment, hill climbing and other plausible heuristics will lead parties toward the median voter.[26] A much harder problem is the so-called paradox of voting. Why do citizens bother to vote? Though not a true paradox, it is a major anomaly for the RC program; as Fiorina once asked, "Is this the paradox that ate rational choice theory?" (1990, p. 334).

The core of the problem is straightforward. In large electorates, it is very unlikely that any one voter will be pivotal. Therefore, if voting is costly, it is not rational to turn out: the costs will exceed the expected gains. (The expected gains equal the chance that one's vote will cause one's preferred party to win times the value of that victory.) Hence, rational citizens will not participate in elections. Yet turnout in stable democracies is substantial; ergo, it is an anomaly. (Although this statement of the problem is decision theoretic, careful game-theoretic models of participation show that essentially the same anomaly plagues strategic models as well—see Palfrey and Rosenthal 1985.) Furthermore, the difference between the models' predictions and the data is huge. No fancy statistical tests are needed; the discrepancy passes the occular test. One needs only to look at standard RC models of turnout, whether decision or game theoretic, to see that they are not even in the empirical ballpark.

There is, of course, a way to rescue the RC theoretical approach in this domain: simply assume that some people have a taste for voting. A well-known example of this move is Riker and Ordeshook's hypothesis (1968) that citizens have internalized a duty to vote, which makes staying home costly. There is undoubtedly something to this: for example, most political scientists are probably "political junkies" who enjoy participating in elections. But for sound methodological reasons, many scholars who work in the RC program are uncomfortable with this move: it smacks of "saving the theory". It is well-known that one can predict virtually any behavior (thus making RC theories unfalsifiable) if one adjusts the agents' utility functions at will, so there is a craft norm of being disciplined about preference assumptions.

Bendor, Diermeier, and Ting (BDT 2003) leave preferences alone—voting remains costly in their model—and instead tackle the problem by assuming that voters are boundedly rational. Our view is that RC models of turnout go wrong because they assume that citizens optimize, thus focusing on whether they are pivotal. (In game-theoretic models, this also implies that citizens solve extremely complicated strategic problems: person A must worry about the possible turnout of B, C, and so on, anticipating that his or her peers do likewise. This boggles the mind.) An alternative way of thinking about this choice is that people learn habits (Simon 1947, p. 88), in politics in particular and in life in general. Habits, in turn, are influenced by a person's prior choices: what seemed to work and what did not.

In this context, certain people learn to vote while others develop different propensities. In BDT's model, citizens learn by trial and error, repeating satisfactory actions and avoiding unsatisfactory ones. This is aspiration-based adaptation: a "satisfactory" action is one that generates satisfactory outcomes, determined by comparing payoffs to one's aspiration level in the typical way. (In BDT, aspirations are endogenous, themselves adjusting to experience.)

The main finding of BDT is that agents who adapt via aspiration-based trial and error turn out in substantial numbers. In most runs of the computational model, turnout averages about 50 percent, even in electorates as large as one million. The explanation for this effect is somewhat involved, but an important part of it is that in competitive districts, aspirations typically fall into an intermediate range, so that *losing is dissatisfying* and *winning is gratifying*, regardless of one's individual action. Hence, losing shirkers and winning voters become more likely to vote in the next election: the shirker's action is inhibited while the voter's is reinforced.

Political demography does the rest: in competitive districts there are many losing shirkers and winning voters. Hence, turnout stabilizes at substantial levels.

## THE LATER SIMON: LITTLE USED

All of the preceding lines of research relied on work that Simon had completed by the early 1960s, and most of it centered on work that had been completed by the late 1950s. But although Simon's earlier publications are better known to political scientists than are his later ones, the latter also have significant implications for our discipline. In this section we look at the later Simon. (For an intriguing and unusual effort to apply the later Simon, see Hammond 2002.) His work in this period is different in method as well as substance: it is more empirical (even experimental), and the models are usually solved computationally rather than analytically. I suspect that when political scientists begin to apply this later work, these orientations might reappear in the applications. Indeed, much of what I examine in this section takes the form of verbal theory, knit together partly by empirical generalizations (e.g., of how expert problem solvers perform so well). The author of spare mathematical models of satisficing, as in "A Behavioral Model of Rational Choice" (Simon 1955), moved to much thicker formulations after 1960.

### The Study of Problem Solving

The bulk of Simon's work in cognitive science examined how humans and computers solve problems. Though decision making is part of problem solving—typically an agent selects a solution to a given problem and so makes a choice—the terminological shift reflects a change in focus. Studies of problem solving spend more time on how agents think about problems than on what choice rule (maximizing expected utility, satisficing, minmax regret, etc.) is used at the end of the process, when a solution is finally selected.

Similarly, behavioral decision theory usually examines behavior in relatively simple contexts, whereas the problem-solving literature often studies subjects facing complex tasks. Further, scholars studying problem solving have spent much more time trying to understand performance variation in given problem domains (e.g., chess) and to explain how certain people (experts) become proficient at solving problems in certain domains. The work of Simon and his longtime collaborator, Allen Newell, greatly influenced all these patterns, and all these emphases in the study of problem solving have significance for political science.

## The Newell-Simon Picture in Brief

The satisficing agent of *Models of Man* (Simon 1957) has a spare psychology. Indeed, the only feature that identifies this construction as part of the cognitive revolution rather than the earlier behaviorism (which, by eschewing mentalistic concepts, generated psychological theories that were extremely stark) is the concept of aspiration levels. As a standard that is internal to an organism, an aspiration level is clearly a mentalistic notion, so it would be avoided by good behaviorists. But the more Simon became a cognitive scientist, the more he became convinced that the depiction in *Models of Man* was too spare to answer the questions he had become interested in. To posit that humans satisfice, when they confront hard problems, tells us how they behave in the search process—when they will stop searching and make a choice. This is useful, and Simon retained the idea in his later, more psychologically detailed work. But it says nothing about how agents think about problems in the first place or how they generate potential solutions.

To address these questions, a more detailed picture is required. The following summary is a crude synthesis of *Human Problem Solving* (Newell and Simon 1972) and some of Simon's later work. (The interested reader is urged to look at the summary chapter of Newell and Simon 1972; Simon 1990, 1996; and Simon's entries in *The MIT Encyclopedia of the Cognitive Sciences*, such as 1999b.) Figure 1 is a simple representation of this summary. Simon's verbal summary is worth quoting:

> To solve a problem, a *representation* must be generated, or a preexisting representation accessed. A representation includes (1) a *description* of the given situation, (2) *operators* or actions for changing the situation, and (3) *tests* to determine whether the goal has been achieved. Applying operators creates new situations, and potential applications of all permissible operators define a branching tree of achievable situations, the *problem space*. Problem solving amounts to searching through the problem space for a situation that satisfies the tests for a solution. (Simon 1999b, p. 674; original emphasis)

Note that cognitive constraints can enter anywhere in the above process. Thus, the question that students tend to focus on, "Do people optimize [i.e., at the end of the process]?" reflects a misunderstanding of the role and significance of these constraints. As Simon remarked (1979b, p. 498), a scholar often has a choice. If he or she assumes that the decision maker's

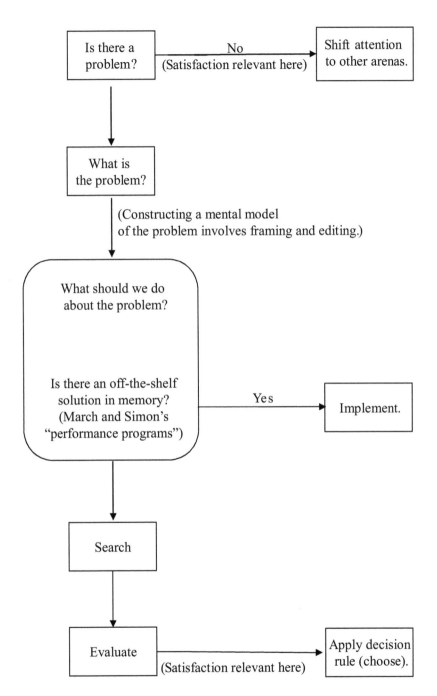

Figure 1. Cognitively constrained decision making.

model of a task is reasonably complex, then it is empirically plausible to assume that at the end of the process, the agent satisfices or uses some other nonoptimizing rule. Alternatively, the scholar can assume that the decision maker optimizes at the end, but then empirical plausibility usually requires assuming that the agent's problem representation is simple. The overarching hypothesis is that when problems are hard, cognitive constraints will pop up *somewhere* in the problem-solving process.

Processes that help humans, especially experts, work around these constraints Simon has called mechanisms for procedural rationality (1990, 1996). Three are worth noting. (The second process, heuristic search, was briefly examined in chapter 1.)

(1)  *Recognition processes.* After many years of experience, experts have stored a great deal of information about their task environments in long-term memory, about both situational patterns ("this is situation type $S_1$") and actions ("in situation type $S_1$, do $A_1$"). For familiar problems, a fundamental trick is to *substitute recognition for search.* It is this—not superhuman calculation or superhuman search speeds—that enables a grand master to play fifty duffers at once.

(2)  *Heuristic search.* When recognition fails, experienced decision makers fall back on search. For reasonably complex problems, search is invariably heuristic. There is no guarantee of optimality.[27] Instead, heuristics make complicated problems manageable: they cut them down to human size and, in the hands of experts, often yield "good" solutions. There are two types of heuristics. If the task is highly (and recognizably) structured, experts use powerful domain-specific heuristics, "drawing upon the structural information to guide search directly to the goal" (Simon 1990, p. 9). If these are unavailable, then decision makers use general but weak heuristics such as satisficing.[28]

(3)  *Problem spaces.* Heuristic search occurs in problem spaces: mental representations of the task at hand. Experts learn to use economical and sometimes powerful problem representations.

Although the above describes a solitary agent, it is straightforward to use this description to develop models of interacting (and conflicting) agents: see, for example, the work on chess. Indeed, much of Simon's work on procedural rationality sprang directly from his study of chess.

## Voters and Jurors, Engineers and Air Traffic Controllers: Hard Problems and Expert Performance

Based only on a reading of the early Simon—in particular, his work on satisficing—one might not have guessed that the later Simon would spend much time studying expertise. In our discipline, a common impression of Simon's approach to decision making is that he assumed that agents "merely" satisfice rather than optimize, which somehow suggests that he focused on mediocre performance. But this is wrong, and significantly so. As the scissors metaphor makes clear, Simon's view of BR was not that humans are dumb but that their problems are often hard.

And in difficult task environments, people's performance can vary enormously. In chess, for example, a few people are grand masters who can pull off amazing feats, such as playing fifty opponents almost simultaneously and beating almost all of them, whereas most of us are duffers. This natural variation cries out for explanation. Thus, starting in the 1960s a new research focus emerged for Simon: how do some decision makers do so well, despite the fact that they too labor under cognitive constraints (Simon and Simon 1962)? Kasparov may satisfice, but he does not *merely* satisfice. What else doe he do, and how does he do it?

Before we inspect Simon's answer, we should note that in his studies of expertise, he selected domains that had two key properties: (a) performance is easy to measure and (b) people vary significantly in problem-solving performance. (Given [a], it is easy to establish [b].) Chess fits these criteria beautifully. There is simply no doubt, for example, that Kasparov is a far better chess player than I am. It is thus no accident that chess became the drosophila of investigations into expertise: that there is a phenomenon to be explained—some people play chess much better than others do—is indisputable.

When we think of using Simon's ideas in political science, we need to think about these properties of the task environment. Do we have good measures of performance? And if so, is there a big variation in performance? Consider, for example, the study of voting. More than most scholars who apply Simon's work, Paul Sniderman has been attentive to the glass-half-full aspect of this work. In several publications (e.g., Sniderman 2000, p. 70) he has referred to "Simon's puzzle": how do voters make sensible decisions in informationally complicated environments? But it is harder to apply the glass-half-full reasoning here than it is in chess. Voters don't compete directly against each other as they do in chess, so we lack a clear performance metric. (For a discussion of the problem of assessing voter competence, see Kuklinski and Quirk 2002.) Or consider foreign-policy makers (Axelrod 1976). Although Axelrod's subjects clearly knew more about foreign-policy issues than does the average citizen, this domain lacks chess's compelling performance data. Indeed, in one study of foreign-policy making, "Across all seven scenarios, experts were only slightly more accurate than one would expect from chance" (Tetlock 1999, p. 351; see also Tetlock 2005, p. 118).

Consequently, in some of these contexts it is an open question whether there are any real experts at all. (I am implicitly defining expertise in terms of problem-solving performance; knowledge in one's head does not suffice.) Indeed, debunking the claims of "experts" to be what they claim is a long-standing tradition in social science, going back to Meehl's classic work (for an overview, see Dawes, Faust, and Meehl 1989; see also Camerer and Johnson 1991). Meehl and his colleagues established beyond a reasonable doubt that in some problem domains, simple linear rules work better than the judgment of specialists. Who knows? Perhaps a Meehlian analysis of foreign-policy making might yield similar results. After all, we don't know whether there truly are experts in that field.[29] Or in voting.

Some readers may object: surely there is demonstrable expertise in some aspects of politics and government. Of course there is, but not in obvious places. (That is, these places may not be obvious to some political scientists, but readers of *Administrative Behavior* would not find them surprising.) Consider specialists in the Army Corps of Engineers or the California Department of Transportation. It is a good bet that civil and structural engineers in those agencies have demonstrable problem-solving expertise. (Living in earthquake country, I take that bet seriously!) Remarkably, the 1989 Loma Prieta earthquake killed fewer than one hundred people, despite its magnitude (about 7.0 on the Richter scale) and location (a metropolitan area with millions of people). A statue honoring these genuine experts, many of whom work in public agencies, should have been erected after Loma Prieta to acknowledge their contribution in helping to avert a catastrophe. (Not long before Loma Prieta, a smaller quake in a much less populated part of Armenia killed thousands. The difference was, I suspect, largely the quality of engineering in the buildings in the two sites.) Or consider cryptographers in the CIA, biostatisticians in the Centers for Disease Control, fighter pilots (especially those landing on carriers; see Wilson 1989) or air traffic controllers anywhere. These roles are much less glamorous than, say, that of the national security advisor, but from a research perspective, they offer the benefit of exhibiting demonstrable expertise. Political scientists have a tendency, perhaps even a systematic bias, to study the exciting and to eschew the mundane. We can learn much from Simon's scientific judgment in this regard. True, chess isn't politics. But drosophila aren't human beings, and geneticists have learned an enormous amount by patiently studying the chromosomes of fruit flies.

Let us briefly examine, then, three very different domains regarding expert performance. (This list is not exhaustive.) In domain 1, all normal human beings perform amazingly well relative to artificial systems. Language acquisition is the paradigmatic example. Despite the "poverty of the stimulus," to use the psycholinguists' phrase, all normal children learn their native language swiftly and easily. Specialists in artificial intelligence are a long way from duplicating this feat. The universality of this performance clearly indicates the operation of powerful evolutionary forces and a resulting hardwired ability. In domain 2, after years of "focused practice" (this term is explicated below), a few human beings develop demonstrable expertise. Chess is the paradigmatic example. The huge performance variation suggests that natural selection has not played a direct role: there may be a "language module" in our brains, but there is no "chess module," and probably none for playing the violin, doing biostatistics competently, or solving cryptographic problems. (Of course, selective forces were undoubtedly important in providing sufficient general intelligence for ordinary humans to learn to do any of those things. But this takes years of concentrated study; learning one's native language doesn't.) In domain 3, assessing expertise is difficult. Indeed, it is unclear whether genuine experts exist in this domain. *Claims* to expertise may, of course, abound. Since the dependent variable hasn't been established, there is no point in trying to explain what may not be there. Instead, debunking may be in order. (Naturally, assigning a particular role—voter, engineer, politician—to a domain is an empirical matter.)

Consider domain 2, in which some people have demonstrable expertise. How do they use it? Recall that Simon's explanation has two main components. First, after years of practice, people specializing in a particular task environment learn to recognize and remember patterns of stimuli— configurations that often recur. Second, specialists learn and remember action rules that are effective for given patterns. Thus, rather than carrying out time-consuming heuristic search, experts simply *recognize* the solution to a given problem. A key implication of these hypotheses is that experts—even the very best, such as chess grand masters—need not have unusual general cognitive capacities such as an exceptionally good long-term memory. Instead, they have accumulated very large and well-organized mental libraries of situations and corresponding actions in a specific domain. These specialized libraries enable them to perform their impressive feats.

A beautiful experiment (Simon and Chase 1973) provides compelling evidence for this claim. Many chess masters seem to have formidable memories: the most spectacular examples are those who can successfully play multiple games blindfolded. Because duffers cannot remember a single game blindfolded, much less many of them, one might hypothesize that chess experts have much better memories than does the average person. Simon and Chase tested this hypothesis versus the competing-mental-libraries hypothesis by comparing the performance of masters and novices in two conditions. First, they gave the subjects a game, played part way through, to study and memorize. The subjects were shown such boards for thirty seconds and then asked to reconstruct them. Experts made far fewer errors than novices did. But then subjects were shown boards on which pieces had been *randomly* arranged—and the error rates of experts and novices converged. Evidently specialists learn to recognize *chunks*: groupings of pieces that often appear in real games. The "magic number seven" (Miller 1956) does indeed constrain the working memory of chess masters as well as novices, but the former don't see or try to remember twenty-seven distinct pieces; they see only six familiar chunks (patterns). This illustrates Simon's theme perfectly: instead of eliminating these fundamental limitations on information processing; experts finesse them. Grand masters *are* boundedly rational, but they have learned how to perform superbly within these constraints.

This takes time, of course. Simon has estimated that a grand master has fifty *thousand* chess chunks stored in long-term memory; learning this many patterns requires playing and practicing for many years—about ten, it seems. Simon and other researchers have suggested that there is a robust empirical regularity here: to perform at a top-notch level in an informationally intensive field—physics, music composition, chess—usually takes about ten years. (This claim is meaningful in domain 2, where demonstrable expertise exists, but not in domain 3.)

With this view of expertise in mind, let us now revisit voting. A major controversy in the field centers on just how competent voters *are* (e.g., Converse 1975; Sniderman, Brody, and Tetlock 1991; Kinder 1998). But before we try to answer Simon's puzzle, it may be sensible to figure out what kind of problem domain voting belongs to. It is clearly not in domain 1. If it were, specialists would not disagree so vehemently about the issue of competence. (Psycholinguists do not disagree about our ability to learn our native tongues: this isn't a "stylized" fact but a plain old-fashioned one.) Further, as Kuklinski and Quirk (2000) argue, it is unlikely that selective forces in our species' evolutionary history put a premium on abilities related to voting in large electorates. Voting might belong to domain 2, but it is less clearly a member of this set than, say, chess: although political parties compete, individual voters do not compete directly with each other; hence, we lack the powerful performance measures available in chess. Another possible performance standard is what fully rational citizens would do, but this one raises the awkward issue that completely rational people with strictly positive costs of voting would rarely vote at all. As with chess, it is desirable to use empirically relevant performance standards rather than those derived from game theory. So we are left with real, interpersonal comparisons. One might use the feasible benchmark of how sophisticated citizens—political scientists, say—with preferences like those of unsophisticated citizens would vote. This, however, may beg a key question: just how able are sophisticated voters? Although it is empirically clear that some citizens have much more political information than others and that they organize this information in more complex ways, do they exhibit demonstrable expertise in the act of voting?

A similar question can be asked about other political roles that, like voting, are necessarily executed by people who do not specialize in that role—for example, jurors. (For a fascinating experimental study of how well jurors carry out their tasks, see Sunstein et al. 2002.) A comparison of the mental processes of jurors and voters, and of the cognitive demands imposed by their roles, would be quite enlightening. Although voters and jurors belong to different subfields in political science, and work in one area rarely cites work in the other, in Simon's theory of problem solving they belong to the same category—amateurs trying to solve cognitively hard problems—so the theory predicts qualitatively similar behaviors. This illustrates how a new research program can reorganize our conceptual landscape and even our academic division of labor.

Finally, it is an open question whether Simon's scissors can be fruitfully deployed on voting, especially in the final election of two-party races, where sophisticated voting is not an issue. Although the decision to *participate* in elections in large jurisdictions is clearly a hard problem, when viewed via game theory's demanding standard, whom one should vote for in the final stage of a two-party race is not. Because voting in primaries or in multiparty races can be significantly more complicated—should I vote sincerely, for my top pick who has no chance, or strategically, for my second choice, who might win?—Simon's scissors might cut more sharply in those contexts. We may find out that such voting does indeed belong to domain 2: after some training and practice, some citizens do learn to vote strategically, while many others, like duffers in chess, don't display this level of expertise.

This conclusion, by leaving us open to charges of elitism, may make us uncomfortable. But such charges would be based on a misunderstanding of modern theories of expertise, especially those influenced by Simon's work. These theories emphasize that expertise is always domain specific (Feltovich, Prietula, and Ericsson 2006). No one is a "general expert"; such beings don't exist. (Clearly, if becoming a first-class expert in information-intensive fields takes about ten years, no one can become proficient in more than a few areas.) This perspective differs sharply from the eighteenth-century elitism of, for example, the founding fathers, not only because it is meritocratic in an effort-based sense (rather than assuming that some people are born more gifted) but also because, in the traditional view, problem-solving ability was an attribute of a *person:* it traveled with him (and the person was always male) from situation to situation. In the modern, domain-specific view, everyone is an amateur in most of life.

## CONCLUSIONS

Whether the work of the early or the late Simon will be more useful for political science in the long run is uncertain. The early Simon is ahead now, and I suspect this trend will continue for some time. His pre-1960 work enjoys significant advantages. In particular, its theoretical structure is much simpler, as a comparison of ("A Behavioral Model of Rational Choice" 1955) and *Human Problem Solving* (Newell and Simon 1972) quickly reveals. Hence, his early formulations are easier to use in formal models of political processes such as elections. In contrast, his subsequent works in cognitive science are like Persian miniatures. The computational models in *Human Problem Solving* are detailed even by the standards of psychology, a field that is more micro than our discipline and that naturally studies individual problem solving and choice in a more fine-grained way.

But these issues are complex. They involve trade-offs between scientifically important criteria—for example, between the theoretical coherence of our discipline and the verisimilitude of particular theories. Scholars who put great weight on coherence prefer research programs whose theories have a simple axiomatic structure and that can unify different empirical domains, even at the cost of some predictive accuracy. Scholars who greatly value verisimilitude are willing to sacrifice theoretical simplicity and generality if they obtain predictive accuracy in return.[30]

Simon himself exhibited different positions on this trade-off over his career. The author of *Models of Man* (1957) evidently placed great store on axiomatically simple formulations, often represented by mathematical models with closed-form solutions. But over time he seemed to put increasing weight on verisimilitude and less on simplicity.[31] Indeed, toward the end of his life he became impatient with mathematical economists who were developing BR models with what he viewed as insufficient attention to the empirical foundations of their premises. His position is clearly expressed in a fascinating exchange with the distinguished game theorist Ariel Rubinstein, who had written a book on formal models of BR and had graciously published parts of Simon's critical letter in an appendix to the book. Simon criticized what he saw as Rubinstein's neglect of empirics:

> Aside from the use you make of the Tversky-Kahneman experiments, for which I applaud you and them, almost the only reference to empirical matters that I detect in your pages is an occasional statement like "a casual observation" and "the phenomenon exhibited here is quite common."
>
> My training in science has installed in me a knee-jerk response to such statements. I ask automatically: "How do you know?" "What evidence can you provide to show that this is true?" Long experience in the natural sciences ... has shown that casual empiricism does not provide a firm foundation for the theories that fit the facts of the real world. Facts do not come from the armchair, but from careful observation and experimentation. (quoted in Rubinstein 1998, p. 188)

In the only face-to-face conversation that I ever had with Simon, a few months before his death, I suggested that political science, as a field with many more macroconcerns than cognitive science, had to be ruthless about the level of detail of its microassumptions. I was fortunate to have encountered the elderly Simon; if the stories are true, the younger man would have gone at me with hammer and tongs, intellectually speaking. Having mellowed, he was gentle. Still, he disagreed firmly and repeated the view that he had articulated in his letter to Rubinstein and elsewhere: what is the scientific value of theories that are empirically inaccurate? I muttered something about trade-offs that varied continuously, rather than being either-or choices, and dropped the matter. Arguing with one's hero is awkward.

But hero worship should not impair our thinking about the design of effective research strategies. I believe that Simon, genius though he was, got it wrong: macrofields such as political science *must* be more ruthless than related microfields (such as cognitive science) about microassumptions. The former face tougher trade-offs than the latter. At the end of the day we are interested in elections, democratic transitions, war and peace—phenomena involving thousands or even millions of people. Theories of such phenomena that are based on individual decision making cannot look like the models in *Human Problem Solving*; the microdetails would overwhelm us.

Simon should have understood this. The author of "The Architecture of Complexity" should have realized—and in other contexts did realize—that when boundedly rational observers of systems aggregate from the micro to the macro, they inevitably delete microdetail. And political scientists are just as boundedly rational as other observers.

Curiously, however, another part of his work—that on expertise—helps to explain why he didn't remember this point when he scolded social scientists who build formal models of BR. As I noted in the beginning of this chapter, for the last forty years of his life he was a cognitive scientist. The axiom of serial processing implies that since he was thinking hard and long about issues in that field, he was *not* thinking much about issues in other fields, such as political science. So he lost touch with his first speciality. As he emphasized in his work with Chase, expertise is domain specific, developed *and maintained* by intense practice and thoughtful reflections about that practice. He had mostly forgotten what it meant to be a political scientist— a natural consequence of cognitive processes that he himself had studied.

This is ironic. In cognitive science, Newell and Simon were famous for their declaration, which they defended throughout their careers, that information processing is quasi independent of the neighboring microfield of neurophysiology. As in-principle reductionists but pragmatic holists, they argued that cognitive science could progress rapidly if cognitive scientists were to temporarily ignore how the brain implements mental processes (Newell and Simon 1972, pp. 875–76). And they had good reason to worry that their new field would get bogged down in neurophysiological details. The human brain is fantastically complicated (billions of neurons, each connected to thousands of other neurons); understanding the physical mechanisms could easily divert researchers from studying higher-order processes such as problem solving.

Of course, Newell and Simon recognized that *eventually* cognitive science would require neuroscientific foundations. But in 1955 they thought that their infant field would benefit from temporarily waving a flag of quasi independence.[32] Much the same holds for the relation between cognitive and political science. Scholars who use psychologically spare models of decision making, such as satisficing and search, recognize that *eventually* these models need the detailed foundations that only more fine-grained cognitive theories can provide. But temporarily asserting quasi independence might be a smart research heuristic. What matters in the macro models of interest to political scientists are only a few properties of human problem solving and choice (Simon 1987), such as seriality in high-order thinking.

Exactly how those properties are produced is something that we might want to set aside for now—along with the detailed information-processing models that would provide the answers. If we are lucky, this academic division of labor will work as well for us as the Newell-Simon approach worked for cognitive science.

CHAPTER 3

# Satisficing

*A* Pretty *Good Heuristic*

JONATHAN BENDOR, SUNIL KUMAR,
AND DAVID A. SIEGEL

## INTRODUCTION

Despite the able efforts of peacemakers (Samuels, Stich, and Bishop 2002; Samuels and Stich 2004; Samuels, Stich, and Fuacher 2004) and a fascinating and unusual "adversarial collaboration" (Mellers, Hertwig, and Kahneman 2001), the debate about rationality marches on at a good clip (e.g., Evans and Over 1996; Chase, Hertwig, and Gigerenzer 1998; Stanovich 1999; Stanovich and West 2000; Gigerenzer and Selten 2001; Gilovich and Griffin 2002; Kahneman 2002; Gigerenzer 2004; Hertwig and Todd 2004; Hertwig and Ortmann 2005). A key part of the debate that has emerged over the last decade is the argument between the heuristics-and-biases camp (e.g., Kahneman, Slovic, and Tversky 1982; Gilovich, Griffin, and Kahneman 2002) and the fast-and-frugal heuristics approach (Gigerenzer and Goldstein 1996; Gigerenzer, Todd, and the ABC Group 1999; Gigerenzer 2001). Crudely put, the former approach emphasizes how badly humans sometimes perform on judgmental and decision-making tasks, while the latter emphasizes that even simple, hence cognitively feasible, heuristics can produce very good results. Each school exhibits important nuances, but we think these are their central tendencies. This debate has been reproduced, on a smaller scale, in political science. (See Lupia 1994 for the optimistic view; Kuklinski and Quirk 2000, for the pessimistic; and Lau and Redlawsk 2001 and Tetlock 2005, p. 119–120, for empirically nuanced positions.)

Given this sharp disagreement, it is intriguing that the two sides claim a common intellectual ancestor: the idea of bounded rationality (Simon 1955,

1956, 1957).[1] Of course, closer inspection might reveal that there is nothing much going on here: it's just a type of intellectual evolution (descent with modification), in which an earlier idea was sufficiently broad to allow for the evolution of quite different descendants. But something deeper may be at work. Central to both approaches, as their names indicate, is the notion of *heuristics*. Roughly speaking (neither approach defines the idea precisely), a heuristic is a rule of thumb that makes complex problems cognitively manageable.[2]

But although both sides regard heuristics as a key concept, they have focused on different *effects* of these rules: the heuristics-and-biases group has emphasized when they go awry; the fast-and-frugal group, on when they do well.

A central thesis of this chapter is that by examining only one type of effect, each side has departed from the core of Simon's vision of bounded rationality. In effect, each approach has highlighted only one side of the *same* coin (or mind).[3] In contrast, Simon's theory of bounded rationality is inherently dualistic regarding human rationality and the power of our problem-solving tools: both our strengths and our weaknesses are central to our understanding of our cognitive processes. Central to this dualistic theory are the following claims:

(1) All humans and their mental procedures have information-processing constraints.

(2) When constraints don't bind they don't matter, and the decision maker and/or his or her procedures can perform optimally or nearly so.

(3) When constraints do bind they matter: they affect both decision processes and outcomes.

(4) These constraints are manifest in some problem contexts but not in others; in real ecologies people invariably encounter some problems that make their constraints show through.[4]

(5) Although humans aren't fully rational, they are adaptively so: problem-solving methods that work well today tend to be used more tomorrow; those that work badly tend to be used less often.

Simon used a homely metaphor to summarize his perspective: "Human rational behavior ... is shaped by a scissors whose two blades are the structure of task environments and the computational capabilities of the actor" (1990, p. 7). In this dualistic view, the cup is *both* half empty *and* half full: it depends on whether a task environment's complexity exceeds a decision maker's information-processing constraints. Hence, Simon's ideas cannot be easily categorized as optimistic or pessimistic.[5] When criticizing economists' theories of choice, he tended to emphasize cognitive constraints: what humans can't do.[6] But when he studied problem solving by experts (e.g., Simon and Chase 1973) without contrasting it to what an idealized fully rational agent would do, he tended to emphasize the positive, adaptive qualities of their performances.

Unfortunately, maintaining this dualism seems to be a difficult mental feat. We suspect that it is an unstable position: like a ball perched precariously on a sharp mountaintop, one tends to slide down toward either the optimistic or the pessimistic orientations.

Our goal is not only to help make peace between the heuristics-and-biases and the fast-and-frugal camps. We want to do more than that: we want to help *unite* them, on the basis of Simon's Janus-faced vision of bounded rationality in general and of heuristics in particular.[7]

Making peace suggests that the two approaches are consistent. We think this is so and that a rigorous analysis, using the work of Samuels, Stich, and Faucher (2004) as a template, would reveal that their *core* claims—not their (rhetorically useful) inflated ones—are compatible.[8] But more than consistency is involved: the two schools' common intellectual ancestor demands that they be *combined*. This is not an optional exercise: under Simon's view of bounded rationality, we have an accurate understanding of our cognitive faculties if *and only if* we understand when and why they work well *and* when and why they work poorly. (This may sound banal, but the history of the "rationality wars" suggests that implementing Simon's perspective isn't easy.) Either side by itself—just good performances or only bad ones—is radically incomplete. We learn just as much about cognitive capacities by studying our failures as by studying our successes: no more, no less.

Consider a simple analogy: exams and students. To accurately estimate how well students understand course material, one should include a mix of easy and hard questions. Students with a deep understanding of the material will be able to answer all the easy questions and many of the hard ones; those with only a superficial understanding may be able to handle many of the easy questions but few of the hard. A test with only easy problems is deficient because it doesn't reveal variation in the cognitive capacities of students. (Stanovich and West have made this point more than an analogy. Their studies, summarized in Stanovich 1999, show that typical heuristics-and-biases questions constitute exactly such tests for most subject pools: subjects who score higher on standard intelligence tests are more likely to give normatively appropriate answers.)

We believe that persuading the two groups that they should form one approach can be done by showing in detail how Simon's original dualistic vision can be carried out. We do this by analyzing the strengths *and* the weaknesses of one heuristic, arguably the most important one in the bounded rationality research program: satisficing.[9]

## THE HEURISTIC OF SATISFICING

Simon primarily saw satisficing as part of a descriptive theory of choice (Simon 1956, p. 104). But he often called it a theory of intendedly rational behavior (1957, p. 196), and philosophers have pointed out that any such theory has a normative component (Byron 2004). Specifically, Simon argued that satisficing has two performance attributes that make it a sensible heuristic for humans facing complex problems. First, as already noted, it is adapted to decision makers' *inner* environment—to their cognitive and informational constraints. Second, satisficing is adapted to their *outer* environment: it often performs "reasonably" well. Hence, Simon never justified satisficing by arguing that it is equivalent (even in the ultralong run) to optimizing,[10] contrary to what some heuristics-and-biases scholars have suggested.[11] Of course, if a problem is *sufficiently* easy, then satisficing might well converge to optimizing, just as more general cognitive limitations (e.g., working memory's storage capacity) might be invisible for some problems but "show through" (Simon 1996) when people confront more demanding ones. All this is perfectly consistent with Simon's scissors.[12]

Given these scissors, a serious study of the satisficing heuristic should carry out the following tasks:

(1) Find out when, if ever, it is optimal, and in what sense (e.g., is it optimal only in the long run?).

(2) Identify the class of problems for which it is suboptimal, and explain why it is suboptimal in that class.

(3) For problems for which it is suboptimal, figure out whether it is nonetheless a "pretty good" heuristic and, if so, in what sense of that important-but-vague phrase.

This chapter undertakes a study of satisficing along these lines. Regarding (1), we show, in the decision-theoretic contexts that Simon studied, that except for a class of problems that are well matched to the heuristic, satisficing does *not* converge to optimal behavior even in the long run. Regarding (2), we explain why the satisficing heuristic is suboptimal when it confronts difficult problems, such as those outside the above "well-matched" set. Regarding (3), we show that even when problems are ill matched to the heuristic, satisficing has several desirable properties, thus confirming Simon's guesses about the heuristic's merits. Importantly, these accomplishments do *not* presume unidimensional payoffs: as Simon had argued, satisficing, like other noncompensatory rules (Hogarth 1987), *does not require that decision makers have utility functions.*[13] Because making trade-offs can be mentally challenging (Slovic 1995; Bettman, Luce, and Payne 1998), this feature of satisficing makes it a more flexible decision-making mechanism than optimizing, which fussily insists that it will go to work only if other parts of the mind give it preference inputs that are unidimensional. One might say that satisficing is a "simple heuristic that makes us *robust.*"

### EXPLORATION VERSUS PERSISTENCE IN THE SEARCH FOR ALTERNATIVES

Decision making under uncertainty often involves a tension between exploration and persistence (March 1991). On the one hand, if an agent does not know which action is best, then he or she will usually need to explore the problem space in order to find it. So restlessness—a willingness to search—can be invaluable. On the other hand, restlessness may also imply that the agent cannot settle down on the optimal action (what we call persistence) once he or she has stumbled across it.

Our focus is on how this tension plays out in a variety of settings. We find that problems may be classified as either well or ill matched to the satisficing heuristic. The former are those where satisficing can resolve the tension between exploration and persistence; the heuristic is *preadapted* to problems in this set. Satisficing cannot resolve this tension optimally if problems fall into the ill-matched set. This section and the next focus on elaborating this distinction; the last section analyzes what happens when satisficing encounters problems for which it is ill matched.

### THE SIMPLEST CONTEXT: SINGLE AGENT, UNIDIMENSIONAL PAYOFFS, EXOGENOUS ASPIRATIONS

To begin, suppose a single agent has four alternatives: $w$, $x$, $y$, and $z$. Option $w$ delivers poor payoffs with certainty, $x$ gives poor or fair payoffs, $y$'s are either fair or good, and $z$ yields good ones with certainty. The agent doesn't know any of this and proceeds by satisficing. Suppose first that the agent's aspiration level were exogenously set to equal a fair payoff. In that case, while a poor outcome could trigger search, the other two outcomes would not, potentially causing the agent to be satisfied with the suboptimal $y$, if it were tried before $z$. In contrast, now suppose the agent's aspiration level were set to good. Only option $z$ would satisfy him or her every time, since $y$ sometimes yields a fair payoff, and the optimal result obtains.

This example suggests that under satisficing, what the agent codes as satisfactory—his or her aspiration level—can strongly affect the outcome. This is in part due to a deterministic feature of satisficing: if the agent's aspiration level is set to fair, he or she will regard fair outcomes as satisfactory *with certainty*. (If this were not true—if the agent could be dissatisfied with fair payoffs—then he or she could not get permanently stuck on the suboptimal $y$.) This point leads directly to our assumption about satisficing.

(A1)   *Each agent, i, has an aspiration level, $a_i$, such that for all t and all histories leading up to t, (i) if $\pi_{i,t} \geq a_i$ then the agent satisfices—that is, uses the same action in $t + 1$—and (ii) there is an $\epsilon > 0$ such that if $\pi_{i,t} < a_i$, then he or she searches for a new option in $t + 1$ with a probability of at least $\epsilon$.*

(A1) formalizes this chapter's key ideas: satisficing (part [i]) and search triggered by dissatisfaction (part [ii]). Extending (A1) to multiattribute choice problems is straightforward: one replaces $a_i$ and $\pi_{i,t}$ by the vectors $(a_{i,1}, \ldots, a_{i,k})$ and $(\pi_{i,1}, \ldots, \pi_{i,k})$, satisficing occurs when $\pi_{i,j} \geq a_{i,j}$ for all $j = 1, \ldots, k$, and so forth. This is completely consistent with Simon's verbal theory: "Aspiration levels provide a computational mechanism for satisficing. An alternative satisfices if it meets aspirations along all dimensions" (Simon 1996, p. 30).

Returning to our $w, x, y, z$ example, we see that a decision maker using a rule that satisfies (A1) with sufficiently demanding requirements can avoid becoming permanently trapped in a suboptimal action. Specifically, an aspiration level above "fair" and below "good" provides just the right mix of exploration and persistence to yield optimality, eventually.

But achieving this mix is not always easy. Consider a second example, often called the "two-armed bandit problem." A gambler is playing a slot machine with two arms. The left arm pays off with some fixed probability $p < 1$; the right, with a fixed probability $q$. If the machine pays off, then it yields a set amount of money; otherwise it gives the customer nothing. Suppose $p > q$, so the left arm is the optimal one.

However, counter to the first example, here the optimal action sometimes fails and yields nothing. This distinction is important, so we clarify it with the following definition: *An action or outcome (consisting of a vector of actions) is* perfect *if it gives each player his/her maximum feasible payoff, $\overline{\pi}$, with certainty. Otherwise it is* imperfect.

The restlessness that was so useful in avoiding a suboptimal result in the first example is the gambler's downfall in this one: the imperfection of the optimal arm and the dynamic generated by (A1) conspire to prevent the securing of the optimal result. Indeed, decision problems in which all actions are imperfect are generally much harder for satisficing rules.

To see this more precisely, suppose that the agent's adaptive rule obeys (A1). Let us focus first on solving the exploration problem, so let the agent's aspiration level equal $\overline{\pi}$, the maximum payoff. Then it is easily established that no suboptimal action is stable, *and even the optimal action is stable if and only if it is perfect*. Hence, conquering the exploration problem via demanding aspirations makes it impossible for satisficing-type rules to stabilize on the optimal action when that alternative is imperfect (as it is likely to be in the real world). Exploration is maximized at the expense of persistence.[14]

Now focus on solving the persistence problem: let the agent's aspiration equal the situation's *minimal* payoff, $\underline{\pi}$. Then it is to easy to show that agents with this aspiration level and who adapt via (A1) master the art of persistence too well: the lack of exploration ensures that *all* actions—suboptimal as well as optimal—are stable. Persistence is maximized at the expense of exploration.

Since neither extreme navigates successfully between the Scylla of insufficient exploration and the Charybdis of insufficient persistence, one naturally wonders whether satisficing based on intermediate aspirations would lead eventually to optimization. This guess is half-right: it holds for some but not all choice settings. Our first proposition draws a bright line between these two sets. We define this bright line now.

DEFINITION 1.    *The decision problem of a weakly generic situation is* well matched *to the satisficing heuristic if the minimum payoff of the optimal action exceeds the minimum payoff of every other action. All other problems are called* ill matched *to the heuristic.*

For examples of each type, consider the following problems. In problem 1 action $x$ is equally likely to give 0 or dollars 2; action $y$ gives dollar 1 or dollars 2 with equal odds. This belongs to the well-matched class because $y$'s minimum exceeds $x$'s. In problem 2 each action delivers 0 or dollar 1.

This is ill matched: the actions' minimum payoffs are the same. (For brevity's sake, we will sometimes call problems that are well matched to satisficing "easy" and the others, "hard.")

The content of proposition 1 justifies definition 1's partitioning of problems into those that are well and ill matched to satisficing and reveals why problems that satisfy the definition's payoff property merit the honorific of "well matched."

PROPOSITION 1.   *If the agent uses an adaptive rule that satisfies (A1), then there exists an aspiration level such that the optimal action is the unique stable state if* and only *if the problem is well matched to the heuristic.*

Simon's scissors cut sharply here: they show that any claims of the universal convergence of adaptive behavior to optimization are incorrect. Even though the environment here is stationary—the decision maker faces the same choice problem over and over—convergence to optimality is not guaranteed. Indeed, it will not happen when a satisficing agent faces problems that are ill matched to that heuristic.[15]

A corollary about ill-matched problems follows immediately from proposition 1: if the optimal action *is* absorbing, then so is some other action. Hence, the process might get trapped in a suboptimal state. Equivalently, if no suboptimal action is absorbing, then neither is the best one.[16]

The distinction between well- and ill-matched problems also helps us understand how satisficing rules handle risky choices—roughly speaking, those in which the best option sometimes produces low payoffs. For example, suppose the agent faces a two-armed bandit problem where one alternative is risky (yielding either $x$ or $y > x$) and the other is riskless (yielding $z$ for sure). The problem is more interesting if neither alternative dominates the other, so assume that $x < z < y$. If the riskless option is optimal, then the problem is well matched to the heuristic: any exogenous aspiration in $(x, z)$ ensures that the agent will always be satisfied by the optimal arm, yet sometimes the suboptimal arm will be dissatisfying and hence will trigger search. But *if the risky arm is optimal, then the problem is ill matched to satisficing.* In that case, no aspiration level can address both exploration and persistence optimally: either both options are always satisfying or neither is.

This example, which illustrates the difficulties that satisficing rules have when confronted by risky-yet-optimal alternatives, generalizes considerably, to a class of choice situations involving a type of risk. In the following results (which follow directly from definition 1), let random variable $X$ be called *more spread out* than random variable $Y$ if $\min(X) < \min(Y)$ and $\max(X) > \max(Y)$.

PROPOSITION 2.   *Suppose the agent uses an adaptive rule that satisfies (A1).*

  (i)   *If the optimal action's payoffs are more spread out than some other action's, then the problem is ill matched to the satisficing heuristic.*

  (ii)  *If the optimal action's payoffs are less spread out than all other actions' payoffs, then the problem is well matched to the satisficing heuristic.*

These results indicate that satisficing-type rules have an interesting bias: they are preadapted to settings in which the optimal action is relatively riskless, in the above sense, but maladapted to environments in which it is relatively risky.

## SATISFICING'S SENSIBLE FEATURES

Having established, per Simon's scissors, that satisficing's performance is intimately tied to problem difficulty, we still must explore what happens when a problem is ill matched to the heuristic, when it does not converge to optimizing in the limit. That this can happen would not have surprised Simon, whose justification of satisficing was not based on claims about its long-run properties. Instead, he thought that it was sensible in the *short run* to use this heuristic when faced with complex problems because it satisfies criteria of both the inner and outer environments: it's feasible for real human beings and often performs "pretty well."

Hence, in this section we study satisficing's sensible properties, in the context of hard problems. Because we want to focus sharply on satisficing's basic properties, we simplify the choice environment by examining bandit problems. As noted earlier, these typically have only two payoffs: one low, normalized to zero, the other high ($h > 0$). Classical statistical decision theories and psychological learning theories often called outcomes in such settings either a "success" (a payoff of $h$) or a "failure" (a payoff of 0).[17] This coding fits naturally with our earlier use of these terms. In particular, we show elsewhere (Bendor, Kumar, and Siegel 2004) that aspirations that adjust to experience will, for a large set of adjustment rules, end up in the $(0, h)$ interval. As a result, decision makers adapting via (A2) and facing binary payoffs will learn to see the zero payoff as a failure; similarly, a payoff of $h$ will be perceived as a success. Therefore, we have chosen not to analyze aspirations explicitly in this section, which makes the model much more tractable. Instead, in this section we assume satisficing directly via (A2):

(A2)   *If $\pi_t = h$, then the agent uses the same action in $t + 1$ that he or she used in $t$. If $\pi_t = 0$, then the agent searches (switches to the other action) with probability $\theta > 0$.*

$1 - \theta$ represents the level of inertia—that is, the chance that a player retries her action in the face of failure. As such, it codifies the exploration-persistence tradeoff: higher values of $\theta$ correspond to more exploration.[18]

The next result, proposition 3, analyzes the relation between an action's failure probability, denoted by $f_i$ for action $i$ ($i = 1, 2$), and the probability that the agent will use it at any specific time, denoted by $p_{i,t}$.

PROPOSITION 3.   *Suppose that (A2) holds. The agent has two actions and the start is neutral: $p_{1,0} = p_{2,0}$.*

  (i)   *For all $t > 0$, $p_{1,t} > p_{2,t}$ if and only if $f_1 < f_2$.*
  (ii)   *For all $t > 0$, $\frac{\partial p_{i,t}}{\partial f_i} \leq 0$; except for a few knife-edge cases the inequality is strict.*

Proposition 3 brings into sharp relief what we mean by "sensible" or "good" performance. A decision rule in this context is sensible if its probabilities of choosing different actions are at all times ordered exactly as the actions' success rates are ranked (part [i]), and if an action degrades, then the rule chooses that action less often (part [ii]).[19] These are, we believe, eminently reasonable requirements: they codify what we intuitively desire from any adaptive rule. With these criteria we see that satisficing performs rather well in the short run in the two-action setting.[20] Simon's conjecture was correct.

Mostly, anyway. Other criteria of good performance could be invoked. For example, we could require that the probability of choosing the optimal arm increase monotonically over time—a kind of probabilistic hill climbing. We now give an example to show that this property does not always hold under the conditions of proposition 3. Suppose that $\theta$, the probability of searching given dissatisfaction, equals 1. Let the chance that arm 1 fails be 80 percent, while arm 2 always fails. Then, given a neutral start, $p_{1,1} = 0.6$, but the probability of selecting arm 1, the optimal action, in period 2 is only 0.52. Although the optimal arm continues to be more likely to be chosen than is the inferior one (property [i] of proposition 3 holds), improvement isn't monotonic over time. Why not?

The culprit is a property that we encountered earlier in this chapter: the agent is too restless. He/she reacts too strongly—too definitively—to failure. Since even the better action fails most of the time and since the agent searches with a probability of 1 in the event of failure, with a chance of 0.8 the decision maker, after trying arm 1, will switch to the inferior one. That implies that he might be more likely to select arm 1 tomorrow if he *isn't* likely to select it today. And indeed that is so. Consider, for example, two biased starts: in one the agent is completely disposed to trying arm 1, so $p_{1,0} = 1$, whereas in the other he has the opposite inclinations, so $p_{1,0} = 0$. Which produces a higher probability of trying the optimal action in the next period? The answer is that doing the right thing in period 1 is much more likely if the decision maker was initially inclined toward the *wrong* action![21] The high failure probabilities combine with the agent's determination to search ($\theta = 1$) to produce an oscillatory pattern. This is inconsistent with monotonic probabilistic improvement. (See Bendor, Kumar, and Siegel 2004 for conditions that are necessary and sufficient for this performance criterion to hold.)

This dynamical aspect of (restless) satisficing is analogous to the behavior of an underdamped spring, as in driving a car with bad shock absorbers over a bumpy road.[22] It exemplifies a general feature of dynamical systems. We suspect that many kinds of feedback-driven adaptation, not just satisficing, exhibit this oscillation when agents are very restless in response to failure. (This idea is examined more closely in chapter 5.)

CONCLUSIONS

Simple heuristics *can* make us smart—sometimes, in certain environments. In other environments a simple heuristic may perform badly (Margolis 2000). This dualism is a direct implication of Simon's scissors. A mental procedure such as satisficing is well matched to certain problem contexts. It may match some so well that it is optimal in those. But other problem contexts will reveal a rule's weaknesses. As we have seen, certain problems are ill matched to satisficing because they show up the crudeness of the heuristic's *discriminatory abilities*. In a two-armed bandit, both arms sometimes fail to pay off: their minimal payoffs are the same. This presents problems for satisficing because this heuristic is partly failure driven (search if payoffs don't meet aspirations), and in bandit problems *all* the actions fail sooner or later. Satisficing is too crude to be optimal even in the long run for such problems because it cannot cleanly distinguish what is suboptimal from what is *imperfect yet optimal*.

Scholars who are pessimistic about heuristics probably find these remarks agreeable. So let us shift the emphasis: our results also show that when satisficing fails to optimize in the long run, it often works "reasonably" well. For example, in stationary bandit problems, satisficing often leads a decision maker to select actions in accord with their underlying (and unknown) success probabilities. Further, even in nonstationary contexts, satisficing gets the choice problem right more than half the time provided that the environment doesn't change too rapidly (Bendor, Kumar, and Siegel 2004). And in these latter contexts, we do not know in general which algorithms *are* optimal, so being able to deploy a pretty-good-though-probably-suboptimal heuristic is quite useful. Scholars who are optimistic about heuristics probably find most of *these* remarks agreeable.

But the point is not to make both pessimists and optimists happy by offering results to each side. Instead, this chapter illustrates a method for investigating mental procedures, such as heuristics, that we think is consistent with the central argument of bounded rationality, as summarized in our five claims at the beginning of the chapter. Together, these claims imply that we should expect *all* widely used heuristics to exhibit both pluses and minuses.[23] Hence, we believe that these premises suggest that instead of trying to uncover *only* the biases produced by a heuristic or *only* the good performances, we should routinely look for both. And we mean "routinely" literally. Such routines would implement the dualism inherent in the bounded rationality research program. Further, we believe, based on the history of the rationality debates, that maintaining a dualistic approach is cognitively difficult. Without the aid of a routinized search for the strengths and weaknesses of a heuristic, people will tend to slide into the more stable states of optimism or pessimism.

Some students of decision making might contend that claim 5—problem-solving methods that work well today tend to be used more tomorrow, while those that work badly tend to be used less—or its evolutionary variants (e.g., evolution selects for mental modules that perform well) imply that our mental procedures are optimal.[24] This conclusion is often buttressed by the repertoire perspective: a specific procedure may be suboptimal in some problem contexts, but our minds contain a large toolkit of procedures and also methods for recognizing different types of problems and assigning them to different, domain-specific procedures, and it is this complete *ensemble* of procedures, including problem coding and solution assigning, that is optimal. Although we think the repertoire perspective is important and intriguing, we believe that the hypothesis that learning rules such as those sketched out by claim 5 produce optimal ensembles is too optimistic. Indeed, we think that this hypothesis is inconsistent with the rest of bounded rationality's central argument. This argument is recursive: it can be applied to claim 5 itself. Claim 5 describes a learning heuristic, similar to Thorndike's Law of Effect. This learning heuristic could be the subject of the kind of analysis used in this chapter. *Such an analysis would show* (see, e.g., Bendor and Kumar 2005) *that this learning heuristic is itself suboptimal in some (ecologically realistic) problem contexts.* It is *not* guaranteed to produce optimal rule ensembles. Of course, by the same central argument of bounded rationality, we expect these rule ensembles often to perform "reasonably well." But "pretty good" isn't best. Best is, well, best: there's nothing better. That's a very demanding standard. Why should we expect the mental procedures of a hominid that evolved in a specific evolutionary environment (Stanovich and West 2000; Linden 2007), in a specific solar system located in a specific galaxy, to satisfy grand optimality criteria that hold for all times and places?

That aspiration is unrealistic.

## CHAPTER 4

# A Model of Muddling Through

JONATHAN BENDOR

Old theories in political science rarely die; they usually just fade away. This has been incrementalism's fate. Even in hindsight, this was a curious end to incrementalism's intellectual trajectory. The basic ideas of "muddling through" were described in extremely well-known publications (Lindblom 1959, 1965; Braybrooke and Lindblom 1963): the 1959 essay has been reprinted in about forty anthologies (see Lindblom 1979, p. 524); the two books are classics. And the theory also received its share of pointed criticisms (e.g., Arrow 1964; Boulding 1964; Dror 1964; Etzioni 1967; Schulman 1975; Goodin and Waldner 1979; Lustick 1980).

Yet neither Lindblom nor his critics were able to carry the day. In particular, Lindblom's claims about incrementalism's effectiveness have been neither established nor refuted. Consider, for example, the controversy over the benefits of changing the status quo only by small steps. Critics raised concerns that "incrementalism would tend to neglect *basic* social innovations because it focused on the short run and sought no more than limited variations from past policies" (Etzioni 1967, p. 387, emphasis in the original); that it would reinforce "the pro-inertia and anti-innovation forces prevalent in all human organizations" (Dror 1964, p. 155); and that it embodied "a complacent acceptance of our imperfections" (Arrow 1964, p. 588). Lindblom (1964, 1979) replied to this criticism, but it is fair to say that the arguments on both sides, though reasonable, are less than compelling. Thus, what is perhaps the best known of Lindblom's claims—that making small policy changes is often superior to making radical ones—retains the status of a conjecture over thirty years after it was first proposed.

Or consider the issue of context. As Lustick reminded us, the debate over incremental versus synoptic strategies should focus on their relative effectiveness in different decisional contexts, for it is unlikely that one is better than the other in all circumstances (1980, p. 342). Braybrooke and Lindblom took this point seriously, devoting an entire chapter to analyzing the match between strategies and their appropriate contexts (1963, pp. 61–79). But the status of these matches—which are themselves hypotheses— remains unclear. Braybrooke and Lindblom asserted that incrementalism is suited to situations involving low understanding (p. 78). However, Lustick hypothesized that incrementalism's relative utility "must be discounted to the extent that the complexity of an organization's task environment is non-decomposable" (1980, p. 346). In their criticism of this essay, Knott and Miller interpreted Lustick's claim to be inconsistent with Braybrooke and Lindblom's:

> The more a problem is characterized by nondecomposable complexity, the less likely it is that all the interaction effects among the parts of the problem will be understood. The less understood these complexities, then the more likely it is that synoptic decision making will fail, since synoptic decision making assumes a comprehensive theory capable of accounting for every relevant factor and interaction effect. Clearly, it is in the presence of little-understood complexities that incrementalism has a comparative advantage over synoptic decision making, if ever. (1981, pp. 725–26)

Lustick's reply in the same volume continued the debate but did not resolve it. (That the debate did not seem to be making progress may be indicated by the fact that one finds fewer articles on incrementalism in the rest of the 1980s.) This is a frustrating and unsatisfactory state of affairs. Three and a half decades after the publication of "The Science of 'Muddling Through,' " we should have a better grasp of the theory's central propositions.

My premise here is that this stagnation occurred primarily because both sides of the debate relied exclusively on informal reasoning.[1] It is very difficult to analyze a strategy's effectiveness without the aid of formal models, either closed-form mathematical models or computer simulations.

The advantages of formalization can be seen by comparing the intellectual history of incrementalism with that of satisficing. Even in ordinary English, the heuristic of satisficing is simpler than the heuristics of disjointed incrementalism. And as the work of Simon (1955, 1956) and others (e.g., Cyert and March 1963; Karandikar et al. 1998; Bendor, Mookherjee, and Ray 2006; Bendor, Kumar, and Siegel 2009; chapter 3 of this volume) indicates, it is quite straightforward to build formal models of satisficing.

Accordingly, this heuristic is alive and well in current research on decision making. The contrast with the theory of incrementalism is sharp.

I shall try to ameliorate this problem by formalizing some of Lindblom's central ideas. The relation between the ensuing formal model and the literary theory of incrementalism is complex. On the one hand, the model is not a full-blown formalization of the complete strategy of muddling through. That would be much too ambitious. As Arrow noted in his review of Braybrooke and Lindblom's *Strategy of Decision,* "It is not easy to give a simple statement of their strategy, since, by its nature, it is intended to be flexible and adaptable to differing situations" (1964, p. 586). Something hard to describe must be hard to formalize as well.

On the other hand, it turns out that in some respects the model is *more* complex than the informal theory: to construct a logically complete model, one must make assumptions about the nature of judgment in decision making (e.g., does judgment vary over time?), about which Lindblom's original formulation said little or nothing. The number and variety of these assumptions (stationary versus nonstationary judgment, weak versus strong competence) indicate that in contrast to Arrow's remark, in some ways the informal theory was too simple, not too complex.

Indeed, by forcing us to confront questions that may have gone completely unnoticed in an informal analysis, formalization can provide a major intellectual benefit. More generally, the formal model can clarify the overall logic of Lindblom's informal theory—not only by filling in holes but also by making obvious exactly which claims are assumptions and which are derivations and by rigorously showing which consequences do indeed flow from basic assumptions.

With the help of a mathematical model, we can discover when a local or incremental search for alternatives is superior and when a bolder approach is better and how improved understanding affects different search strategies—the two controversies noted above. We are also able to address several less controversial claims of Lindblom's theory: the purported advantages of seriality (repeated attacks on the same policy problem) and of redundancy (multiple decision makers working on the same problem). So that we do not give away too many of the results, let us note here only that incremental search fares much worse than Lindblom had conjectured and that although seriality and redundancy are often advantageous, under certain well-specified (and not implausible) conditions, even they do more harm than good. Thus, the critiques of Etzioni and Dror were well focused: though the uncontroversial claims about seriality and redundancy do not hold up perfectly, they do better than the controversial one of local search.

Because Lindblom's ideas are complex, we take a leaf from his own work and build the models up serially, beginning with the simplest. Thus, the first model focuses on a single decision maker. Even in this simple context the model captures several essential features of muddling through: it represents a myopic and fallible decision maker, who chooses by comparing the status quo to a few new options and who tries in a groping manner to choose objectively better policies over poorer ones. The decision maker's limitations are partially eased by the fact that he or she makes and remakes policy in "an endless series"—seriality. Following his emphasis on amelioration, the decision maker is interested only in whether a new alternative is better than the status quo.

A simple choice process underlies all the models. In period 1, the decision maker confronts a pair of alternatives and then chooses one and discards the other. In period 2, a new policy is generated from the available pool of possibilities. It is then pitted against the status quo (last period's selection), and a new choice is made. This process is repeated indefinitely. We can then study the impacts of seriality (how the status quo's quality evolves over time), of incremental versus bolder search and how changes in policy understanding affect the relative utility of incrementalism.

The second model introduces an important institutional method for compensating for the cognitive limits on individual agents: multiple decision makers. Here we are interested only in the sheer effect of increasing the number of choosers. We suppress conflict of interest, assuming instead that the decision makers have the same objectives. The third model allows the decision makers to have conflicting objectives and analyzes the path of the status quo over time. We thus build, step by step, to a dynamic model of boundedly rational decision makers who have incomplete information and conflicting objectives. Such models are so complex that simulation is often the only way to solve them (e.g., Bendor and Moe 1986; Kollman, Miller, and Page 1992). Nevertheless, even the last model yields analytical results.

## THE MODEL WITH A SINGLE DECISION MAKER

I shall first focus on a single decision maker. Though this player makes a sequence of choices over time, we do not endow him or her with information about the pool of feasible alternatives.[2] Instead, the player simply compares the status quo with a new alternative, chooses the one that he or she believes is superior, and throws away the other one. The process then continues to the next period.

## Concepts and Notation

In every period, a new option is generated from a set of possible types of policies; this generation process will be described shortly. For all but one of the results, the set of possible policies is assumed to be unbounded; the different kinds will be labeled $\{\ldots, -2, -1, 0, 1, 2, \ldots\}$. Higher integers denote objectively better policies.

Because the decision maker does not know for sure which type of policy he or she is confronting, the key property is competence—the probability of choosing the better of two options. In theories such as Condorcet jury models in which decision makers must make one pairwise choice (guilty or not guilty), with neither option considered the status quo, a decision maker's competence is completely described by a single number, a probability that is presumably in $(\frac{1}{2}, 1)$. Naturally, if jurors err more often than fair coins would, the latter could replace the former. Hence, it may be reasonable to use this assumption in a first pass at the problem. Moreover, the assumption has the virtue of simplicity.

It is not, however, a fully satisfactory cornerstone for a model of muddling through for three reasons. First, the decision maker makes choices repeatedly. Hence, time could matter: the decision maker might, for example, become more competent with experience. Second, because there can be more than two types of alternatives, the probability of being correct might vary across pairs of options. Suppose, for example, that alternatives can be poor, mediocre, or excellent. Compare two choice situations: in one, the status quo is poor and the new proposal is mediocre; in the other, the status quo is poor and the new option is excellent. Presumably it is easier to determine that (an objectively) excellent new alternative is better than (an objectively) poor status quo than it is to conclude that a mediocre option is better than a poor one. Hence, if the status quo is poor, the probability of accepting a new alternative that is excellent should exceed the probability of accepting one that is only fair.[3] Third, models that are status quo–based must allow for the possibility that the status quo policy has special significance in the decision maker's mind. This possibility is especially salient in theories of incrementalism, and it has received empirical support (Kahneman, Knetsch, and Thaler 1990).[4]

Given these theoretical and empirical considerations, one might hesitate before embracing the assumption that the chance that a decision maker will choose correctly always exceeds one-half. And any model of muddling through must make *some* kind of assumption about the effects of time, multiple pairwise comparisons, and the status quo. These three issues will be addressed by the definitions that follow. First, some useful notation. Let $p_{i,j;t}$ denote the chance that given a status quo $i$ and a new option $j$ at date $t$, the decision maker will choose the new alternative over the status quo. (Here, $i$ and $j$ are integers denoting different quality policies.) If $j > i$, then $j$ is objectively better than $i$; hence in this case $p_{i,j;t}$ is the probability of deciding correctly. If $j < i$, then $p_{i,j;t}$ is the probability of making the wrong choice between $i$ and $j$.

As a benchmark case, let us reconsider the simple Condorcet assumption that the decision maker has a constant probability, $p > \frac{1}{2}$, of being correct. If we embed this premise in a model of muddling through, we see that it makes certain presuppositions about the effect of time and of multiple pairwise comparisons and about the impact of the status quo.

*Time*    Because in the benchmark case a decision maker has a fixed and constant probability of being correct, judgment is unaffected by time. I shall say that judgment is *stationary* if $p_{i,j;t}$ equals some constant $p_{i,j}$ in every period, for every status quo $i$ and every new option $j$. Judgment is nonstationary if this condition does not hold. Some of the results that follow allow for nonstationary judgment.

*The Effect of the Status Quo*    In the benchmark case, the status quo is unimportant. In a comparison between alternatives $i$ and $j$ (where, e.g., $i < j$), it does not matter which one is the status quo: the chance of choosing correctly is $p$ in either case. Thus, $p_{i,j;t}$, which (given $i < j$) is the probability of correctly discarding an inferior status quo, equals $1 - p_{i,j;t}$, the chance of correctly keeping a superior status quo. Hence, the decision maker is not biased for or against the status quo. I shall say that judgment exhibits *status quo insensitivity* if $p_{i,j;t} = 1 - p_{i,j;t}$. Judgment is sensitive to the status quo if this condition does not hold.

*Multiple Pairwise Comparisons*    With the benchmark assumption, every choice situation looks the same in that, in any given period, the chance of moving "forward" (selecting an objectively better new alternative) is always $p$ and the chance of moving "backward" (accepting an objectively poorer one) is always $1 - p$. In this, the simplest possible type of judgment, the agent can only distinguish superior from inferior options. Any new option that is better than the status quo looks like any other superior new alternative, no matter how much better one or the other is. Similarly, all inferior new proposals look the same. I shall call this property *crude judgment*. Formally, judgment is crude if for all $h < i$ and all $j < k$, $p_{h,i;t} = p_{j,k;t}$. As a mnemonic notation, when judgment is crude, $f_t$ denotes the probability of moving forward from any status quo in period $t$ (i.e., $f_t \equiv p_{i,j;t}$ for all $i < j$) and $b_t$ denotes the probability of moving backward from any status quo in that period ($b_t \equiv p_{i,j;t}$ for all $i > j$).

Judgment can be more discriminating if it is *homogeneous* across status quos. Judgment that is homogeneous can take into account the difference between a new proposal that is vastly better than the status quo and one that is only a bit better and likewise between a new option that is barely worse than the status quo and one that is a disaster. Formally, all that is required for judgment to be homogeneous across status quos is that for all integers $k$, $p_{i,i+k,t} = p_{j,j+k;t}$. For example, the probability of going from a status quo of 10 to a new alternative worth 12 must equal the chance of going from a status quo of 20 to a new alternative worth 22. Note that crudeness is a special case of homogeneity; judgment can be homogeneous without being crude. (For example, if judgment is homogeneous but not crude, then the chance of discarding a status quo of 10 would be greater if the new alternative were worth 80 than it would be if the new option were worth only 12. Crude judgment requires that those probabilities be the same.) Judgment is heterogeneous if it is not homogeneous.

Finally, one must make some assumption about the agent's competence. A decision maker will be called *weakly competent* if for every status quo $i$, (1) $p_{i,j;t} \leq p_{i,j+1;t}$ for all $j$, $j + 1 \neq i$, and (2) $p_{i,i+1;t} > p_{i,i-1;t}$. Property (1) is a natural assumption: inferior new options cannot be more likely to induce an agent to discard a given status quo than are better ones. Together, the two properties ensure that even a weakly competent decision maker is more likely to discard a fixed status quo in favor of a superior new policy than to do so in favor of an inferior new one.

A decision maker will be called *strongly competent* if property (1) holds and, in addition, $p_{i,j;t} > \frac{1}{2}$ whenever $i < j$ and $p_{i,j;t} < \frac{1}{2}$ whenever $i > j$. Thus, a strongly competent agent has more than a 50 percent chance of selecting correctly, no matter what kind of choice situation confronts her. Weak competence does not imply this level of accuracy in all situations. Clearly, then, strong competence implies weak competence, but the converse does not hold.

Weak competence permits the decision maker to be biased toward (or, less plausibly, against) the status quo. Thus, even if $i < j$, $p_{i,j;t}$ need not exceed one-half; indeed, it could be close to zero. If it is, then the player is very biased toward the status quo. Hence, if $h < i$, then $p_{i,h;t}$ must be still closer to zero. Even in the family of models of bounded rationality, weak competence is a very mild assumption. A bias favoring the status quo (plus even a bit of evaluative accuracy) should imply that the decision maker is more likely to move forward from the status quo than backward. In contrast, strong competence implies that the decision maker cannot be predisposed in favor of the status quo if the latter is inferior to a new proposal.

We can now step back and summarize. The benchmark assumption imported from Condorcet jury models that posit that a decision maker has a constant probability of choosing correctly presumes strong competence and judgment that is crude, stationary, and insensitive to the status quo.

The generation of new options is assumed throughout to be Markovian: given a status quo of $i$ in period $t$, there is a conditional probability $q_{i,j;t+1}$ of producing alternative $j$ as the challenger in the following period. Consistent with Lindblom's position on the constraints on policy design, I assume that in any period only finitely many new types of policies are possible.

As with choice, one must specify the effect of time and the role of the status quo. We will say that alternative generation is *stationary* if $q_{i,j;t} = q_{i,j}$. Thus, the chance of generating an option of type $j$ in period $t$ depends only on the status quo in $t - 1$, not on the date, $t$. If this condition does not hold, the process is nonstationary.

Policy generation is *homogeneous* across status quos if $q_{i,i+k;t} = q_{j,j+k;t}$ for all $k$. That is, if in a given period one normalized two different status quos to zero, the new alternatives associated with each would be distributed identically. Thus, taking into account the value of each status quo, the distribution of new alternatives looks the same. If this condition does not hold, the process is heterogeneous. (The best-known Markov chains, such as random walks, are both stationary and homogeneous.)

Because the process must start somehow, there is a dummy period—period o—in which an alternative is generated, but because there is nothing to compare it to, this option is the agent's de facto choice in this period. Thus, in period o there is an initial probability vector, $[q_1, \ldots, q_M]$, that specifies the unconditional probability of drawing any type of alternative. The real choice process begins in earnest in period 1.

Combined, the assumptions about generation and choice create a Markov chain. A probability transition matrix for this type of stochastic process, with stationary judgment and policy generation, is shown in figure 2.

The state of this chain at any date is the status quo policy. Let $V_t$ denote the value or quality of the status quo policy in period $t$. Because $V_t$ is produced by a choice process marked by errors in judgment and randomness in the generation of alternatives, it is a random variable. $V_t$ is our primary object of interest.

The System in Period $t + 1$

| | | $i$ | | $J$ | | $k$ | |
|---|---|---|---|---|---|---|---|
| | | ... | | ... | | ... | ... |
| | $i$ | $1 - \sum_{h\neq i} q_{i,h}\, p_{i,h}$ | | $q_{i,j}\, p_{i,j}$ | | $q_{i,k}\, p_{i,k}$ | |
| The System in Period $t$ | $j$ | $q_{j,i}\, p_{j,i}$ | | $1 - \sum_{h\neq j} q_{j,h}\, p_{j,h}$ | | $q_{j,k}\, p_{j,k}$ | |
| | $k$ | $q_{k,i}\, p_{k,i}$ | | $q_{k,j}\, p_{k,j}$ | | $1 - \sum_{h\neq k} q_{k,h}\, p_{k,h}$ | |

Note: $q_{i,j}$ denotes the probability that given a status quo of $i$, a new alternative, $j$, will be generated; $p$, denotes the probability that given a status quo of $i$, the decision maker will choose the new option, $j$.

Figure 2. Probability transition matrix.

To prevent the triviality of the choice process from becoming stuck in a particular state, such as $q_{i,i} = 1$ (so that if the agent ever chooses a policy of type $i$, no other type of alternative will ever be produced for consideration), I shall assume that it is always possible to generate new options that are better or worse than the status quo. Hence, in every period there are strictly positive integers $j$ and $k$ such that for every status quo $i$, $q_{i,i-j;t} > 0$ and $q_{i,i+k;t} > 0$.

Given these definitions, alternative generation must be either homogeneous or heterogeneous and either stationary or nonstationary. Similarly, judgment must be either sensitive or insensitive to the status quo, either homogeneous or heterogeneous, and either stationary or nonstationary. Because each pair is exhaustive, whenever one of the following results obtains under either (say) stationary or nonstationary alternative generation, then the statement of that result will not refer explicitly to that property.

*The Effect of Seriality*

We are now ready to examine how the quality of the status quo changes as time unfolds. The first theorem focuses attention exclusively on how the agent's judgment affects the value of the status quo policy over time. This is achieved by holding the generation of new alternatives probabilistically constant: it is assumed that the distribution of new proposals is independent of the status quo (i.e., for all $i$ and $j$, $q_{i,j;t}$ just equals some constant $q_{j;t}$)

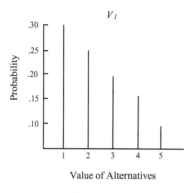 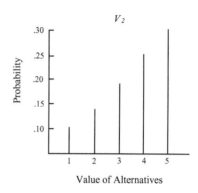

Figure 3. Stochastic improvement.

and that this process is stationary: $q_{j;t} = q_j$. (Given stationarity and the premise, used throughout, that only finitely many new alternatives are possible in any period, it follows that the policy space in theorem 1 must be finite. The possible alternatives are labeled $\{1, \ldots, M\}$, where higher numbers denote superior policies.)

Given these assumptions, the chance of generating, say, the best possible alternative, $M$, remains the same at every date. Consequently, if $p(V_t = M)$ increases over time, we know that all of this improvement in $V_t$ must be attributed to the agent's evaluations of alternatives, not to how those options are produced. Theorem 1 shows that this improvement is marked.

To analyze the change in $V_t$, we use the criterion of stochastic improvement. Suppose $V$ takes on values $\{m, \ldots, M\}$ and $V'$ has support $\{m', \ldots M'\}$, where $m \leq m'$ and $\{M \leq M'\}$. Then $V'$ is stochastically better than $V$ if $p(V' > v) > p(V > v)$ for $v = m, \ldots, M' - 1$. If $p(V' > v) \geq p(V > v)$ for $v = m, \ldots, M' - 1$, we say that $V'$ is weakly stochastically better than $V$. Thus, stochastic improvement means shifting probability weights toward better outcomes (see figure 3).

It is worth pointing out that if $V_t$ gets stochastically better over time, then $E[V_t]$ must also rise. The converse does not hold: the average may increase without stochastic improvement. Hence the latter is a stronger criterion than the former.

THEOREM 1.  *If the decision maker is weakly competent, judgment is crude and stationary, and the distribution of new proposals is independent of the status quo, then $V_t$ improves stochastically with time.*

Note that no structure has been imposed on the distribution of new options: they may be multipeaked, sharply skewed toward poor options, or the like. Regardless of the shape of this distribution, more time means stochastically superior decisions. Of course, the fact that the decision maker is weakly competent is absolutely essential to this result. Indeed, given the other assumptions, weak competence is a necessary condition for $V_t$ to improve stochastically. If the decision maker were more likely to select an inferior new policy over the status quo than to choose a superior new one over the status quo, then $V_t$ would *degrade* stochastically over time.

Theorem 1 is useful as a baseline because we know that none of $V_t$'s improvement is due to decision makers' learning how to generate better options. But empirically we do expect the processes of generation and evaluation to be intertwined. Accordingly, all of the results that follow allow for this interdependence.

The process postulated for the next theorem is extremely general: as before, the agent need only be weakly competent, but now his or her judgment can be either stationary or nonstationary. Moreover, the generation of alternatives may depend upon the status quo, and in ways that may be both nonstationary and heterogeneous.

The bad news is that we must give up the criterion of stochastic improvement in this environment. The culprit is partly the generality of alternative generation. A policy process that admits such generality in how new policies are produced can reach new (heretofore unreached) lows, as well as new highs. (For example, suppose that the status quo in period o must equal zero. New proposals in period $t$ equal $\pm 1$ of period $t$'s status quo. This random walk can take on a value of $-1$ in period 1, of $-2$ in period 2, and so on; hence, stochastic improvement is unattainable.) We therefore fall back on $V$'s average value, $E[V_t]$, as the normative standard.

The most important assumption of theorem 2 is that the new proposals, given a status quo, are on average at least as good as that status quo. To write this concisely, let $Q_{t+1}$, a random variable, denote the value of new alternatives at the start of period $t + 1$; hence, $E[Q_{t+1}|V_t = i]$ denotes the average value of new alternatives in $t + 1$, given that $i$ was the status quo in $t$.

THEOREM 2.    *If $E[Q_{t+1}|V_t = i] \geq i$ and the decision maker is weakly competent, then $E[V_t]$ increases monotonically over time.*

This result can be well understood if we consider an important special case. If $Q_t$ is symmetrically distributed around the status quo, the burden of improving the average is thrown entirely on the evaluation process. Because the agent is weakly competent, he or she must be at least a tad more likely to select better options than to err by embracing weaker ones. Given that $Q_t$ is symmetric around the status quo $i$, the decision maker is more likely to progress to an alternative worth $i + k$ than to fall back to its symmetric twin worth $i - k$. Hence the average must rise.[5]

## Incremental Search and Uncertainty

Lindblom's best-known claim is that making incremental policy changes is generally superior to making radical ones. To investigate this idea systematically we must first define a *less incremental* policy process. The intuitive idea is that incrementalism involves small policy changes; nonincrementalism, large ones (Braybrooke and Lindblom 1963, p. 62). For example, financing public schools completely by vouchers would be a big change; requiring that schools teach fourth-grade mathematics in a new way would be a more incremental change.

More abstractly, let there be a (possibly multidimensional) space that describes the key programmatic characteristics of different alternatives. A search strategy is simply a point in this space. The distance, $d$, between the status quo and the search strategy measures the degree of nonincrementalism in programmatic change: the bigger $d$ is, the more radical the attempted change. Thus, consistent with at least some versions of the informal theory, "we draw no sharp line between the incremental and the nonincremental; the difference is one of degree" (Braybrooke and Lindblom 1963, p. 64).

Nonincremental search does not imply comprehensiveness, in Lindblom's sense of the word. Only a small part of the policy space is investigated. And, as usual, only a finite number of new options are possible in any period. Indeed, one could assume that *fewer* options are generated when policy design is radical. Thus, nonincremental search, though not local, remains limited.

Having defined incremental search, we can now examine Lindblom's argument about how it is supposed to affect policy processes.[6] The basic idea is that nonlocal policy design creates more uncertainty. However, more than one type of uncertainty might arise. First, some distributions of new options will be riskier—hence more uncertain—than other distributions. (A precise definition of this idea will be given shortly.) Second, the evaluation of any new alternative may involve more or less uncertainty.

Lindblom's argument seems to have been that *both* kinds of uncertainty increase as search becomes less incremental: the distribution of new policies becomes riskier, and evaluation becomes "noisier." On the first interpretation, Braybrooke and Lindblom have said that decision makers who muddle through "focus on the increments by which the social states that might result from alternative policies differ from the status quo. . . . They are focused on incremental alteration of existing social states" (1963, p. 85). And the assertions of Etzioni (1967) and Dror (1964) about the pro-inertia tendencies of incrementalism indicate that critics, too, see the strategy's cautious approach as producing new alternatives that are not innovative (i.e., not risky). On the second interpretation, Braybrooke and Lindblom asserted that "an analyst is often without adequate information, theory, or any other organized way of dealing systematically with nonincremental alternatives" (1963, p. 89); hence, such options are typically "unpredictable in their consequences" (Lindblom 1959, p. 85).

Whatever were the original beliefs of Lindblom and his critics, both assumptions seem plausible. It is reasonable to assume that incremental search is less risky: nonincremental search should be more likely than more local search to generate extremely bad and extremely good policies. And evaluating a new alternative that differs greatly from the status quo should be ridden with more uncertainty than evaluating one that differs only marginally. What is less clear is whether these two types of uncertainty affect policy outcomes in the same direction. The next set of results shows that in fact, *they do not*. Whereas noisier evaluation has the harmful effect anticipated by Lindblom, the increased riskiness of a set of new options is benign. To show these distinct effects, we analyze them separately. In the first subsection that follows, it is assumed that less incremental search makes the distribution of new alternatives riskier; there is no effect on evaluation. In the second subsection, it is assumed that more radical search makes evaluation noisier; there is no effect on the distribution of new options. Finally, we examine interactions between the distribution of new policies and their evaluation.

*Incrementalism and Risky Sets of New Options*    Recall that $Q_t$, a random variable, denotes the value of new alternatives at the beginning of period $t$. Let us then use $Q_t'$ to refer to the value of new options generated by a more radical search process; $Q_t$, the value of those generated by a more incremental process. (Hereafter, variables with prime superscripts will always be associated with less incremental search.) The key assumption is that the distribution of $Q_t'$ is riskier or more uncertain than is that of $Q_t$. Roughly speaking, this means that $Q_t'$ is more spread out than is $Q_t$.[7]

To avoid rigging the results either for or against incrementalism, I also assume that $Q_t'$ is a *mean-preserving spread* of $Q_t$: $E[Q_t | V_t = x_0] = E[Q_t' | V_t' = x_0]$. Thus, given the same status quo, the two processes produce

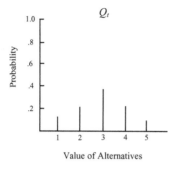 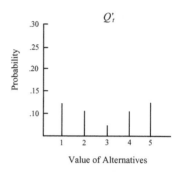

Note: Theorem 3 assumes that $Q'_t$, the riskier distribu tion of new alternatives, is produced by bolder search and that $Q_t$, the less risky distribution, is produced by more incremental search.

Figure 4.  Mean-preserving spread.

equally valuable new alternatives, on average (see figure 4 for an example). No other constraint is placed on the shapes of the two distributions.

We can now compare the average status quo under more versus less incremental processes. Perhaps surprisingly, it turns out that *this* kind of uncertainty is helpful.

THEOREM 3.   *Suppose search in the $V'_t$ process is less incremental than in the $V_t$ process, alternative generation in each process is homogeneous, the decision maker is weakly competent, and his judgment is crude. Then $E[V'_t] > E[V_t]$, for all t.*

The intuitive explanation for this result follows. Suppose both incremental and nonincremental search generate new alternatives that are distributed symmetrically around the status quo, with the latter's distribution being more dispersed. Consider the extreme case in which there is no chance of moving backward ($b = 0$) by mistakenly accepting a new option that is inferior to the status quo. Hence, the greater downside risk associated with radical search is irrelevant; all that matters is its greater upside risk, which is pure gravy. More generally, since a weakly competent agent is always more likely to move forward than backward ($f > b$), the upside risk outweighs the downside.

It is straightforward to show that given the theorem's assumptions about alternative generation and judgment, weak competence is *necessary* for radical search to be superior to incremental search. If the decision maker is weakly incompetent ($f < b$), then bold search hurts rather than helps.

Thus, theorem 3 identifies the *minimal level of rationality* that is required in order to reject Lindblom's conjecture.[8]

*Incrementalism and Noisy Evaluation*    Here I assume that more radical search affects evaluation but not the distribution of new alternatives. To make this idea precise, let us posit that the *perceived* value of new options, $\tilde{Q}_t$ equals their true value plus an error term that represents "noise." Naturally, the decision maker who perceives the new option to be better than the status quo selects it; otherwise, he keeps the status quo.[9] Thus $\tilde{Q}_t = Q_t + \theta_t$, where $\theta_t$ is any continuous random variable. (We don't need $E[\theta_t]$ to equal zero: the evaluation of new options may be biased. Hence, the decision maker may exhibit a status quo bias.) It is only required that $\theta_t$ can take on both positive and negative values, so that the agent can either overestimate or underestimate the value of a new alternative.

As the notation suggests, $\theta_t$ may vary over time; consequently, the decision maker's judgment may be nonstationary. Within the same period, however, the error term produced by a given search strategy is constant. This implies homogeneous judgment.[10] Observe also that this error structure does not imply even weak competence: only one of the attributes of a weakly competent agent, that $p_{i,j;t} \leq p_{i,j+1;t}$, is implied.

Paralleling the previous notation, let $\theta'_t$ denote the noise associated with evaluating a new policy generated by a more radical search process, and let $\theta_t$ denote that associated with evaluating an incrementally generated option. We will say that $\theta'_t$ is *noisier* than $\theta_t$ if it is more spread out—that is, has fatter tails.[11] To keep the comparison clean, we shall assume that $\theta'_t$ is a mean-preserving spread of $\theta_t$. Hence, this definition implies that evaluating radical options is riskier than evaluating more incremental options, with greater riskiness being formally defined in exactly the same way that greater riskiness was defined for the distribution of new alternatives. But here greater riskiness is indeed harmful.

THEOREM 4.    *Suppose that search in the $V'_t$ process is less incremental than in the $V_t$ process and that less incremental search produces noisier evaluation. Alternative generation in each process is identical and is homogeneous. Then $V_t$ is (weakly) stochastically better than $V'_t$, for all t.*

Searching more incrementally yields only weak stochastic improvement because less noisy evaluation does not always imply that judgmental accuracy increases. (Of course, the chance of being right cannot decrease as noise falls.) Whether less noise means greater accuracy depends upon how the error terms, $\theta_t$ and $\theta'_t$, are distributed in relation to the values of the status quo and a given new alternative. The increased noisiness of $\theta'_t$ may occur in a judgmentally irrelevant region. For example, suppose that in every period, $\theta_t$ is distributed uniformly over $[-0.4, 0.4]$ and $\theta'_t$ is uniform over $[-0.8, 0.8]$. Then the judgment is flawless in each process, and $V_t$ and $V'_t$ will be distributed identically in every period.

There are, however, plausible restrictions on the error term that do imply that $V_t$ is stochastically better than $V'_t$ in the normal, strict sense. For example, if both $\theta_t$ and $\theta'_t$ are normally distributed, then judgment in the more incremental process will be strictly better than judgment in the less incremental one. This in turn implies that $V_t$ is stochastically better than $V'_t$ without qualification.

*Interactions*    We now examine interactions between the distribution of new alternatives and their evaluation. The first result focuses on how better policy understanding affects the value of bold search in different policy domains. In their typology of decision contexts, Braybrooke and Lindblom argued that incrementalism is more appropriate in situations in which there is little understanding (1963, p. 78; see also Knott and Miller 1981, p. 726; Braybrooke 1985, p. 926). One kind of understanding is manifested in accurate policy evaluation: the more consistently decision makers choose good alternatives over bad ones, the more understanding they display. Thus, Braybrooke and Lindblom's argument implies that in domains where there is great understanding, decision makers can more easily afford to take the risks of radical change.[12]

As shown by theorem 3, the direct effect of less incremental search is to increase the average quality of the status quo, even when the agent is weakly competent and policy understanding is poor ($f$ is less than one-half and close to $b$). This was unanticipated by Lindblom's argument. However, if Braybrooke and Lindblom's thesis about the match between strategy and policy context is correct, then the *relative* benefits of bolder search should rise as judgmental competence increases. For their specific conjecture about the interaction between incrementalism and policy understanding to be correct, it does not matter whether $E[V'_t] - E[V_t]$ is positive or negative; what matters is how this difference changes as $f$ rises or $b$ falls.

Interestingly enough, although Lindblom's hypothesis about incrementalism's direct effect on policy is invalid under the conditions of theorem 3, his and Braybrooke's conjecture about the *interaction* between incrementalism and policy understanding turns out to be correct.

THEOREM 5.   *If the conditions of theorem 3 hold, then the relative benefits of less incremental search, $E[V'_t] - E[V_t]$, increase as $f$ rises or $b$ falls.*

The intuitive explanation for this result is straightforward. If $f = 1$ and $b = 0$, then the decision maker gets the full benefit of the greater dispersion associated with bold search: superior new options are always chosen; weaker ones, always rejected. If $f = b$, then one is just as likely to move backward as to move forward. Hence, novelty in policy design is irrelevant. As $f$ rises or $b$ falls, more and more of the potential benefits of nonincremental search are realized.[13]

As suggested earlier, it is plausible that searching more radically will increase *both* kinds of uncertainty: the set of new policies will become riskier, and the evaluation of any one alternative will become noisier. The final two results of this section examine contexts in which both effects occur. Because this is analytically more difficult, I must make more restrictive assumptions. Therefore, $\theta$ cannot be any continuous random variable; instead, here it is distributed normally with mean zero and variance $\sigma^2$, and it is stationary as well. The noisiness of evaluation is parameterized by $\sigma^2$. I assume that $\sigma^2$ increases as search becomes more radical. The evaluation of a given new alternative depends upon the distance, $d$, between the status quo and the search strategy that generated the new option. Specifically, $\sigma^2 = d^\alpha$, where $\alpha$ is a positive constant. Thus, $\alpha$ parameterizes how well incremental search strategies reduce evaluation uncertainty, compared to nonincremental strategies. As such, $\alpha$ is the focus of the next result.

As a further simplification, I dichotomize the search strategies. Instead of comparing continuous variation in search, as in theorems 3, 4, and 5, the next two results use categorical comparisons. As a normalization, call the search *incremental* if $d < 1$; call it *nonincremental* if $d > 1$. Propositions 1 and 2 will compare a fixed incremental strategy—a particular $d$ less than 1—with a fixed nonincremental strategy—a $d'$ greater than 1.

Given the nature of this comparison, observe that a very large $\alpha$ represents a world in which being incremental during policy design gives the decision maker a big advantage in the policy evaluation phase, because $\sigma^2 = d^\alpha$ falls toward zero as $\alpha$ increases (since $d < 1$), whereas $\sigma^2 = (d')^\alpha$ blows up (since $d' > 1$).[14] A value of $\alpha$ close to zero represents a world in which incrementalism during policy design gives very small leverage during policy evaluation.

The next result focuses on a fixed pair of search strategies, $d$ and $d'$. Because $d' > d$, the distribution of new alternatives generated by $d'$ is, as before, riskier than the distribution generated by $d$. But now evaluation is affected as well. Specifically, let $Q'_t = Q_t + \theta$, where $\theta$ is distributed normally with mean zero and variance $\sigma^2$, with $\sigma^2 = d^\alpha$.

PROPOSITION I. *Suppose alternative generation is homogeneous and stationary and let $Q'_t$ be riskier than $Q_t$ as described above. The decision maker is strongly competent, with judgment that is homogeneous, stationary, and insensitive to the status quo. Then there exists an $\alpha^* > 0$ such that for all $t$,*

  (i)   *if $\alpha < \alpha^*$ then $E[V_t] < E[V'_t]$;*
  (ii)  *if $\alpha > \alpha^*$ then $E[V_t] > E[V'_t]$.*

A parallel result can be established for increasing degrees of riskiness. Again consider a fixed pair of search strategies: $d < 1$ and $d' > 1$. We represent the effects of local versus nonlocal search as follows. Assume that new alternatives are homogeneously generated and distributed symmetrically around the status quo. For any status quo $x_0$, let $x_0 - m$ denote the smallest value of $Q_t$ and $x_0 + m$ its largest value, where $m$ is a strictly positive integer. Because $d' > d$, we will assume that the smallest and largest values of $Q'_t$ are more extreme than those of $Q_t$. Specifically, let $x_0 - m - k$ denote the smallest value of $Q'_t$, and $x_0 + m + k$ its largest. The parameter $k$ is determined as follows. For any positive $x$, if $x$ isn't an integer, then let $[x]$ be the smallest integer that exceeds $x$; if $x$ is an integer, then $[x] = x$. Then for a fixed $\beta \geq 0$, $k = [d^\beta] - 1$. Thus, if the search is incremental ($d < 1$), then $k = 0$ for any value of $\beta$. If search is nonincremental ($d > 1$) and $\beta$ is strictly positive, then $k$ must be at least 1.

We must also specify how variations in $\beta$ (for a fixed search strategy $d$) and variations in $d$ (for a fixed $\beta$) affect $Q$'s complete distribution, not just its extreme values. Suppose $\beta_1 < \beta_2$. Given a fixed search strategy $d$, let $k_1 = [d^{\beta_1}] - 1$ and $k_2 = [d^{\beta_2}] - 1$. Let $Q_1$ and $Q_2$ be stationary distributions of new alternatives associated with $\beta_1$ and $\beta_2$, respectively. The following assumptions are useful simplifications. If $k_1 = k_2$ due to the rounding occasioned by $[\cdot]$, then $Q_1$ and $Q_2$ are distributed identically. If $k_1 < k_2$, then $Q_2$ is a mean-preserving spread of $Q_1$. Thus, if $\beta$ becomes sufficiently bigger so as to make the biggest and smallest values of $Q$ more extreme, it makes $Q$ itself riskier. If the change in $\beta$ is insufficient to alter $Q$'s extreme values, then the entire distribution of $Q$ remains unchanged as well. In general, then, because the parameter $k$ is weakly increasing in $\beta$, changes in $\beta$ reflect how much dispersion in new options one gets for a given amount of nonlocal search.

We make analogous assumptions about the impact of changes in $d$ on $Q$'s distribution. Suppose for a fixed $\beta$ we have $d_1 < d_2$. If $k_1 = k_2$, then $Q_1$ and $Q_2$ are distributed identically; if $k_1 < k_2$, then $Q_2$ is a mean-preserving spread of $Q_1$.

Finally, to ensure that a riskier distribution puts a nonnegligible increased weight on extreme values, I assume that there is an $\epsilon > 0$ such that for all $Q_t'$ (i.e., no matter how large $k$ is), if $p(Q = x) > 0$, then $= p(Q_t' = x) \geq \epsilon$.

PROPOSITION 2. *Assume the conditions of proposition 1, with a fixed $\alpha$ (hence fixed variance). Suppose the above assumptions on the distributions of $Q$ and $Q'$ hold. Then, apart from an unimportant knife-edge case (see Bendor 1995), there exists a $\beta^* \geq 0$ such that for all $t$,*

(i) *if $\beta \geq \beta^*$ then $E[V_t'] > E[V_t]$;*

(ii) *if $\beta < \beta^*$ then $E[V_t'] < E[V_t]$.*

Propositions 1 and 2 together imply that if making the search more radical simultaneously produces more dispersed alternatives and noisier evaluations, then the parametric space is partitioned into two regions (figure 5). In one region, nonlocal search makes evaluation so difficult that this effect outweighs the benefits of riskier distributions of new alternatives ($\alpha$ is high, $\beta$ is low). In the other region, the balance of effects is the opposite: evaluating unusual programs is not as hard and nonincremental search yields much riskier distributions of options ($\alpha$ is low, $\beta$ is high).

## MANY DECISION MAKERS, SAME GOALS

A single, boundedly rational decision maker can do only so much, even if he or she exploits seriality by repeatedly attacking a problem. As Lindblom pointed out, "One can easily imagine a decision maker who can return in later policy steps to no more than a few of a variety of neglected adverse consequences resulting from an earlier policy step. This possibility points directly to the need for a multiplicity of decision makers and, more than that, to a multiplicity marked by great variety of attitudes and interests, so that no line of adverse consequence fails to come to the attention of some decision maker" (1965, p. 151).

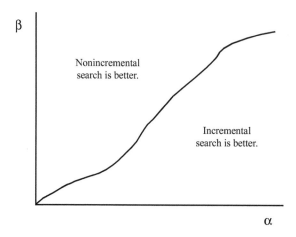

Note: α parameterizes how much the noise in evaluation is reduced by searching incrementally; β parameterizes how much riskier are distributions of new options that are generated by nonincrementalsearch.

Figure 5. The relative superiority of different search strategies.

I shall investigate the effects of a "multiplicity marked by great variety of attitudes and interests" shortly. Here I examine the consequences of sheer size—increasing the number of (independent) decision makers.[15]

Let there be $n \geq 3$ decision makers. The group decides by majority rule, so let $n$ be odd. I shall assume that the decision makers are clones of each other: in every period, given any pair of alternatives, every agent has the same probability of being correct. Because adding decision makers is pointless if individual judgment is perfect, we assume that $p_{i,j;t}$ is always less than 1 for $i < j$ and always exceeds zero for $i > j$. As Condorcet assumed in the static setting of a single choice, the decision makers' judgments are independent (but see Ladha 1992). Judgment is insensitive to the status quo. Finally, each agent is strongly competent. I call this set of assumptions Condorcet conditions.[16]

We are now ready to examine how increasing group size affects the quality of the group's choices in every period. Theorem 6 builds on the classical Condorcet result that in a static setting involving a pair of alternatives, the probability that a majority of the group will make a correct decision increases as more (equally competent and independent) decision makers are added. Though the Condorcet conditions are required, in other respects the result holds under rather general conditions. In particular, neither alternative generation nor judgment has to be stationary; both may change over time, reflecting learning or forgetting.

THEOREM 6.  *Suppose the Condorcet conditions hold and both judgment and alternative generation are homogeneous. Then increases in n make $V_t$ stochastically better, for all t.*

To see the family resemblance between this result and Condorcet's Theorem, focus attention on the best possible policy of any given period. (Because only finitely many new options are possible in every period, for any fixed period a best possible policy exists, even though over time the policy space is unbounded.) Then theorem 6 implies that adding decision makers increases the chance of picking, in any fixed period, the best possible policy of that date. Similarly, adding agents decreases the chance of picking the worst possible alternative in any given period.

### Interactions between Increasing Size and Seriality

It is clear that judgmental accuracy and the quality of new proposals are substitutes. The better the former, the poorer the latter can be in order to achieve a fixed level of improvement in $E[V_t]$. The opposite relation holds as well: if the decision makers have very good judgment, then the new alternatives can generally be inferior, yet the average status quo policy will still improve. And since we know that under the Condorcet conditions, increasing the group's size improves its collective competence, we should be able to offset weak policy generation by making the committee sufficiently large.

This intuition underlies the next result. I have assumed that every distribution of new alternatives has potential in that there is always a chance that something better than the status quo will be produced. Having potential is a very mild condition. (Indeed, it is so weak that it is obviously a *necessary* condition for $E[V_t]$ to rise: if it does not hold, then even flawless evaluation can do no more than cling to the status quo.) Weak as it is, it provides the basis for theorem 7.

THEOREM 7.  *Suppose the Condorcet conditions hold. Both judgment and alternative generation are homogeneous and stationary. Then there exists an $n^*$ such that for all $n > n^*$, $E[V_t]$ is monotonically increasing over time.*

Given the Condorcet conditions, a sufficiently large group ensures that even a feeble alternative generation process provides enough opportunities for improvement, on average. The result holds because, under the classical assumptions, the probability that a majority of the group will vote correctly goes monotonically to 1 as the group grows larger. Then, to see immediately why the theorem is valid, consider the limiting case of an infallible majority. In this circumstance the committee never mistakenly discards a status quo for something inferior, and it always throws out the status quo when something better turns up. Since $Q_t$ has potential, there is always some chance that something better than the status quo will be generated. Thus, in the limit, the average policy must improve. Since the convergence to majoritarian infallibility is monotonic, a sufficiently big $n$ ensures that we can get as close to this ideal situation as is desired.

Naturally, the better the set of new policies, the smaller is the critical value of $n^*$. Consider the important special case of a martingale process: $E[Q_{t+1} | V_t = v] = v$ (March and Olsen 1984). In this case, we know from theorem 2 that $n^* = 0$: if the committee is composed of only one person, $E[V_t]$ will rise steadily.

Of course, there are limits to what even a very large group can do if new alternatives are usually very poor. Thus, one can easily imagine a bad "dual" to theorem 7: for any fixed $n$, no matter how large, and for any fixed individual competency level, no matter how close to 1, there is a sufficiently bad distribution, $Q_{t+1}^\bullet$, such that for all distributions stochastically worse than $Q_{t+1}^\bullet$, policies on average degrade over time.

## Interactions between Increasing Size and Incremental Search

Let us briefly return to the context of theorems 3 and 5, in which search affects only the riskiness of the distribution of new alternatives; evaluation is unaffected. Recall from theorem 5 that in policy domains where evaluation is good, the relative value of nonincremental search, $E[V'_t] - E[V_t]$, is higher than in those where evaluation is poor. We also know that under the Condorcet conditions, increasing $n$ boosts majoritarian competence. Suppose that the Condorcet conditions hold, together with the other assumptions of theorem 5. It then follows as an immediate corollary that the relative benefits of nonlocal search are increasing in $n$.

What is the effect of increasing $n$ if nonlocal search increases *both* kinds of uncertainty, by making the set of new options riskier and evaluation noisier? Suppose the incremental process ($V_t$) and the nonincremental process $V'_t$ have the same number of decision makers. It turns out that as the groups grow larger at the same rate, the scales must eventually tip in favor of radical search.

THEOREM 8.   *Suppose bolder search produces both riskier sets of new options and noisier evaluation. In each process, the Condorcet conditions hold; judgment and alternative generation are homogeneous and stationary in each. Then there exists an $n^*$ such that for all $n > n^*$, $E[V'_t] > E[V_t]$, for all t.*

This result holds because of Condorcet's Theorem. His result tells us that in both incremental and nonincremental processes, majoritarian judgment gets arbitrarily close to infallibility as $n$ rises. This implies that any difference in the evaluative abilities of the two groups must get arbitrarily small, thus allowing the other effect of nonlocal search—the greater dispersion of new alternatives—to predominate. Hence, if one thoroughly applies Lindblom's prescription of redundancy, one should *not* follow his recommendation about incremental search.

It is instructive to compare theorem 8 to figure 5, which illustrates the fact that given a single agent, the parametric space divides into two regions: in one, incrementalism is superior; in the other, nonincremental search.[17] In the world of figure 5, if $\beta$ is high and $\alpha$ is low, then $n^\dagger 0$. Clearly the value of $n^\dagger$ must rise when we move into the region where $\alpha$ is high and $\beta$ is low, for there local search is better than nonlocal when there is only one agent. Nevertheless, even in this region, radical search is better for sufficiently big $n$.

## MANY DECISION MAKERS, DIFFERENT GOALS

Here the agents are specialists, trained to worry about either, say, the environmental impacts or the economic consequences of a policy, but not both. Thus, we continue to have $n$ (odd) decision makers who use majority rule, but now they represent $n$ different policy dimensions. An alternative is a point in this $n$-dimensional space. For example, if there are three dimensions, the status quo might be $(1, -4, 5)$, and a new option might be $(3, 1, -88)$. If the Condorcet conditions hold, then the group is more likely to pick the new option (even though it is dreadful on dimension 3) than to keep the status quo. The motivational interpretation of this behavior is that each decision maker cares only about his or her speciality; the cognitive interpretation, consistent with bounded rationality, is that each one perceives an alternative only in terms of its effect on the relevant speciality. Thus, when presented with a choice between $(1, -4, 2)$ and $(3, 1, -88)$, the specialist on dimension 1 sees only a choice between an option worth 1 and another worth 3.

This neglect of side effects is a major part of what Braybrooke and Lind-blom call "disjointed" incrementalism.[18] In *The Intelligence of Democracy,* Lindblom calls this behavior parametric adjustment, noting that it is the simplest kind of partisan mutual adjustment. "It is adjustment by a decision maker who does not try to control another decision maker, does not defer to the other decision maker, does not allow the possible reactions of the other decision maker to his decision to influence the decision he makes. Whatever adjustment he achieves is due to his accommodating himself to a state of affairs without contemplating what he might do to the other decision maker" (1965, p. 36).

Because specialists myopically ignore the effects of an alternative on other policy areas, one might guess that this disjointed collective behavior creates systematic problems, just as individual-level heuristics cause systematic biases (Kahneman, Slovic, and Tversky 1982; Gilovich, Griffin, and Kahneman 2002). In two of the most serious problems, disjointed incrementalism can cause (1) a majority of departments to recommend a series of proposals that will on average hurt each department or (2) a majority to reject a series of proposals that would on average help every department. Both problems are illustrated by the following example. Given a status quo of $(0, 0, 0)$, six new alternatives are generated with equal probability: $(-5, 1, 1), (1, -5, 1), (1, 1, -5)$ and $(5, -1, -1), (-1, 5, -1),$ $(-1, -1, 5)$.[19] For simplicity, assume initially that the agents evaluate flawlessly. Then the new options of $(-5, 1, 1), (1, -5, 1),$ and $(1, 1, -5)$ will be accepted because in each case, two specialists correctly believe that the new alternative improves their domain. But over time this behavior will hurt everyone: the expected value of these three alternatives, on each dimension, equals $\frac{1}{6}(-5) + \frac{1}{6}(1) + \frac{1}{6}(1) = -\frac{1}{2}$.

In contrast, the options of $(5, -1, -1), (-1, 5, -1),$ and $(-1, -1, 5)$ will be rejected because a majority of the agents correctly believe that the new alternative hurts their domain. But if this is done serially, they will forgo improvement in all three specialities: the expected value of these three alternatives, on each dimension, equals $\frac{1}{6}(5) + \frac{1}{6}(-1) + \frac{1}{6}(-1) = \frac{1}{2}$. Thus, although the average new alternative is just as good as the status quo, over time the average status quo will degrade. Because the worst options are the ones that gain majority approval, seriality will hurt rather than help. This example generalizes to a class of choice situations (theorem 9, part [ii] to come).

Of course, decision makers are fallible, so the degradation will be probabilistic rather than certain. This matters. In the example just given, infallible specialists always make the wrong collective choice; fallible ones sometimes make the right collective choice by mistake. Thus, for this kind of problem, if evaluation becomes more accurate, the average status quo degrades *faster*, so that increasing the number of independent specialists on each dimension only makes matters worse (see theorem 10, part [ii]).

These results are obtained by generalizing the feature of the above example of $(-5, 1, 1), \ldots, (-1, -1, 5)$ that poses difficulties for myopic, majoritarian incrementalism. Recall that in this example, when a new option was better for a majority of the departments, it was much worse for a third. Because this held symmetrically and because all new alternatives were equally likely, this meant that the subset of three options was on average inferior to the status quo for each department—but each option by itself was preferred to the status quo by the majority. Conversely, when a majority preferred the status quo over, say, $(-1, -1, 5)$, the underlying subset of options would have made all departments better off. In both cases what the majority wanted, when decision makers focused their attention on the alternative before them, was inconsistent with what they would have wanted if they had inspected the set of options more broadly. Hence, I call such options *majority inconsistent*. (See Bendor 1995 for a precise definition of majority inconsistency.)

Normatively, majority inconsistency cries out for greater strategic sophistication. The agents should logroll across dimensions by agreeing to reject alternatives that appear to hurt one domain badly and to accept new options that they believe are a big improvement in one dimension. It is unclear, however, whether such sophistication is consistent with disjointed incrementalism.[20]

Theorem 9, part (i), and theorem 10, part (i), identify conditions that ensure that these troublesome choice situations cannot arise. Consequently, both seriality and redundancy improve matters, as Lindblom conjectured. These conditions are the opposite of majority inconsistency: they hold when an option is preferred by a majority of departments and when the option is better, on average, than the status quo for every department. Call this *majority consistency*. (See Bendor 1995 for a precise definition.)[21]

Like theorem 2, which it resembles, theorem 9, part (i), also assumes that the expected value of new options, conditioned on a given status quo, is at least as good as that status quo's value on each dimension.

THEOREM 9.   *Suppose the agents are weakly and equally competent, with crude and independent judgment.*

(i)  *If every dimension's new options are majority consistent and*
     $E[Q(X_{i,t+1})|X_{i,t} = x_i] \geq x_i$ *for all dimensions $i = 1, \ldots, N$, then*
     $E[X_{i,t}]$ *increases monotonically over time, on all dimensions.*

(ii) *If every dimension's new options are majority inconsistent and*
     $E[Q(X_{i,t+1})|X_{i,t} = x_i] \leq x_i$ *for all dimensions $i = 1, \ldots, N$, then*
     $E[X_{i,t}]$ *decreases monotonically over time, on all dimensions.*

A close reading of the theorem reveals that parts (i) and (ii) share an important special case, in which new proposals on average exactly equal the status quo. In this martingale process, everything depends on the pattern of conflict among policy dimensions. If new policies are majority consistent, all is well. But if they are majority inconsistent, then the expected status quo gets worse. Note that this difference between parts (i) and (ii) does not depend on any difference in the decision makers' competence: in each case they are weakly competent. Even if the agents in part (ii) were more competent than those in part (i)—indeed, *even if they were infallible*—the average status quo policy would still degrade. Bounded rationality does matter in part (ii), but it does not take the form of judgmental incompetence. What drives part (ii) is myopia—inattention to an alternative's effects on a different policy domain—combined with a particular pattern of conflict across domains.

Next we reconsider the effect of increasing the number of agents. We do so symmetrically: the organization increases from $n$ to $mn$, where there are $m \geq 3$ (odd) decision makers specializing in each of the $n$ policy dimensions.[22] Here the specialists on a particular policy are considered to constitute meaningful departments: the status quo is kept if a majority of the departments vote to do so. In turn, a department will cast a vote to retain the status quo if a majority of its members vote to do so. Hence the probability that the status quo is kept is the chance that a majority of people in a majority of departments are in favor of so doing.[23]

THEOREM 10.   *Suppose there are n departments, each with m agents. The Condorcet conditions hold. Judgment is crude and stationary; alternative generation is homogeneous and stationary.*

(i)  *If every dimension's new options are majority consistent, then*
     $E[X_{i,t}]$ *is monotonically increasing in m, for all $i = 1, \ldots, n$ and all $t > 0$.*

(ii) *If every dimension's new options are majority inconsistent, then*
     $E[X_{i,t}]$ *is monotonically decreasing in m, for all $i = 1, \ldots, n$ and all $t > 0$.*

Theorem 10, part (i), says that if there is the right pattern of conflict among policy areas, then adding decision makers is benign. The following example gives the intuition. Suppose the six equally likely new options are $(6, 6, -1)$, $(6, -1, 6)$, $(-1, 6, 6)$ and $(-6, -6, 1)$, $(-6, 1, -6)$, $(1, -6, -6)$. Then over the subset of alternatives that improve a majority of dimensions, each department is on average better off; conversely, over the subset that impairs a majority, each department is on average worse off. In such a situation one would like each department to judge as competently as possible, and under the Condorcet conditions making each unit bigger does the trick. The intuition behind part (ii) is the preceding example of $(-5, 1, 1), \ldots, (-1, -1, 5)$.

Finally, we reconsider the effect of making search more or less incremental. For a single decision maker, theorem 3 showed that if search affected only the riskiness of new proposals, making search bolder improved the average status quo policy; theorem 4 showed that if search affected only the noisiness of evaluation, then incremental search was superior. Proposition 3 shows that both of these effects stand up in the multiperson setting, given majority consistency; as usual, majority inconsistency reverses the results.

Analyzing how search affects multiple decision makers who attend to different goals is complicated. To keep the problem tractable, two simplifying assumptions are useful. The first simplification concerns the circumstance in which search affects only the riskiness of options (proposition 3, part [i]). It is assumed that bolder search has a constant effect in each dimension $i$, by multiplying the distance from the status quo to each new option by an integer $c_i > 1$ on each dimension. For example, suppose that given a status quo of $(0, \ldots, 0)$, the more incremental process generates a new proposal of $(x_1, x_2, x_3)$ with probability one-fourth. Then the less incremental process generates an option of $(c_1 x_1, c_2 x_2, c_3 x_3)$, also with probability one-fourth. (The $c_i$ can vary across dimensions.) Since each $c_i$ exceeds 1, $Q_t'(X_i)$, the distribution of values on the $i$th dimension produced by more radical search, is riskier than $Q_t(X_i)$, which is produced by more incremental search. I call $Q_t'$ an *expansion* of $Q_t$. To ensure that $Q_t'$ is a mean-preserving spread of $Q_t$, it is assumed that new policies are martingales: $E[Q(X_{i,t+1}) | X_{i,t} = x_i] = x_i$.

The second simplification pertains to how search affects the evaluation of new options (proposition 3, part [ii]). The process of judgment is black-boxed, and it is simply maintained that less incremental search makes evaluation less reliable: $p_{i,j}' < p_{i,j}$ for $i < j$ and $p_{i,h}' > p_{i,h}$ for $i > h$.

PROPOSITION 3.    *The decision makers are weakly and equally competent; judgments are independent and crude. Alternative generation is homogeneous.*

   (i)    *Suppose bolder search yields riskier distributions of options. New policies are martingales, and $Q'_t$ is an expansion of $Q_t$. If every dimension's new options are majority consistent, then $E[X'_{i,t}] > E[X_{i,t}]$ on every dimension and for every t; if they are majority inconsistent, then $E[X'_{i,t}] < E[X_{i,t}]$ on every dimension and for every t.*

   (ii)    *Suppose bolder search produces less reliable evaluation. If every dimension's new options are majority consistent, then $E[X'_{i,t}] < E[X_{i,t}]$ on every dimension and for every t; if they are majority inconsistent, then $E[X'_{i,t}] > E[X_{i,t}]$ on every dimension and for every t.*

The key to understanding proposition 3 is realizing that it satisfies the conditions of theorem 9. Thus, in proposition 3, part (i), if majority consistency holds, theorem 9 implies that regardless of the degree of incrementalism, the expected status quo improves. Since the average new policy just equals the status quo, credit for improvement must go to the process of collective evaluation, which weights the good part of the set of new options (where the average new option exceeds the status quo) more than the bad part. Given this, the benefit of scaling up the distribution's good part by a positive constant must outweigh the cost of scaling up its bad part by the same amount. So it pays to search more boldly, because doing so yields expanded (scaled-up) distributions. When majority inconsistency holds, theorem 9 says that the average status quo falls, because collective evaluation weights the distribution's bad part more than the good. Given this, scaling up the bad and good parts equally is harmful. Thus, more local search, which produces more compact distributions, is superior.

Part (ii) of proposition 3 is straightforward. Given majority consistency across policy domains, majoritarian tendencies are beneficial for each department. Therefore, one wants departments to be as judgmentally competent as possible. Hence, if nonincremental search makes evaluation less reliable, while leaving the distribution of new options unchanged, then it must be harmful.

But search probably affects both kinds of uncertainty simultaneously. What then can be said about the *net* effect of more versus less incremental search? Again Condorcet comes to the rescue. Recall that theorem 8 showed that when both effects are present, bolder search is superior if the (harmonious) group is sufficiently big. This holds when there is conflict, so long as majority consistency obtains.

PROPOSITION 4. *Suppose bolder search produces both riskier distributions of new options (via expansions) and less reliable evaluation. The Condorcet conditions hold. Judgments are crude and stationary. New policies are martingales, generated by a homogeneous and stationary process. Each department has n agents.*

  (i)   *If every dimension's new options are majority consistent, then there is an $n^*$ such that for all $n > n^*$, $E[X'_{i,t}] > E[X_{i,t}]$ on every dimension and for every t.*

  (ii)  *If every dimension's new options are majority inconsistent, then there is an $n^*$ such that for all $n > n^*$, $E[X'_{i,t}] < E[X_{i,t}]$ on every dimension and for every t.*

Proposition 4, part (i), suggests that there is nothing inherent about conflict that makes nonincremental search less valuable. What does cause problems for bold search is (as part [ii] reveals) a particular *pattern* of conflict: majority inconsistency. When this pattern holds, the more localized that search is, the better.

But part (ii) gives little comfort to Lindblom's theory, for the property that makes incremental search superior yields the *wrong* effect regarding the theory's uncontroversial claims about seriality and redundant agents. When new alternatives were majority inconsistent, theorem 10 reported that adding decision makers *worsened* the average status quo policy and theorem 9 reported that making choices repeatedly did likewise. Hence, the informal theory of disjointed incrementalism faces a dilemma: conditions confirming its uncontroversial claims help to disconfirm its claims about the benefits of local search; and conversely, conditions upholding the conjectures about incremental search yield unexpected results about seriality and redundancy.


CONCLUSION

The results show that when there is a single decision maker or a harmonious group, then working on a problem repeatedly or adding new agents are helpful heuristics, as Lindblom hypothesized. But when there are many decision makers with conflicting goals, even these procedures can back-fire: Lindblom's claims, widely regarded as uncontroversial, turn out to be invalid under the not-implausible circumstance of majority inconsistency.

These results underscore the significance of the relation between *heuristics* and the *biases* they can produce. This line of research, pioneered by Kahneman and Tversky in the context of single-person judgment and choice, has now been extended by Neale and Bazerman (1991) to mul-tiperson strategic settings. Given the recency of these developments, it is unsurprising that the much older theory of incrementalism failed to analyze how the heuristics embodied in muddling through might create systematic problems under certain conditions. But the notion that all heuristics can do so—no matter how useful in certain environments—is now conventional wisdom in behavioral decision theory. Applying this insight to the study of complex political institutions—in particular, understanding how heuristics that are adapted to certain patterns of conflict are maladapted to others—should be an important item on our research agenda.

Consider now the purported benefits of incrementalism's most hotly debated heuristic, local search. Here the informal theory runs into problems even with a single decision maker. Even in this simple setting, the advantages of incrementalism are equivocal. One reason for this lack of clarity is that the original literature did not specify precisely how search affected uncer-tainty. Thus, some conceptual unpacking was required. I've suggested that Lindblom and his critics believed that nonincremental search created two very different kinds of uncertainty: a riskier distribution of new alternatives and noisier estimates of the value of any new alternative. These two kinds of uncertainty affect the average status quo policy in sharply contrasting ways. Theorem 3 and proposition 2 show that greater uncertainty concerning the distribution of new alternatives is beneficial, whereas theorem 4 and proposition 1 show that noisier evaluations are harmful.

These results are analytical: they examine one type of uncertainty while holding the other constant. The problem is that as one searches less locally, both kinds of uncertainty will probably increase, and the *net* effect of these two changes is uncertain. However, theorem 8 tells us that the net effect will favor bold search if there are enough decision makers who, Con-dorcet fashion, produce collectively reliable evaluations. And proposition 4, part (i), tells us that under the same condition (majority consistency) that confirmed the verbal theory's uncontroversial claims, the net effect still favors radical search. Thus, the logic of incrementalism's benefits is either unclear or rejected. Because this is one of my more surprising results, it merits a few concluding remarks.

First, in some policy areas, improving on current solutions may be hard because most of the promising ideas have been tried (Kingdon 1984). Or perhaps the cost of searching rises sharply as policy design becomes more innovative. For either reason, nonincremental search might generate new options with lower expected net value than those produced by local search. (Contrast this with the assumption used here of riskier distributions being mean-preserving spreads of safer ones.) To be sure, simply asserting that nonincremental search produces poorer average new alternatives is not a deep point theoretically: as stated, it is either a theoretical assumption (not a derivation) or an empirical generalization.[24] Nevertheless, if true it would certainly be of great pragmatic importance.

Second, a related empirical claim is that implementing radically new programs is harder than implementing marginally different ones (Braybrooke 1985). In addition to case study evidence (e.g., Derthick 1990), the phenomenon of the "learning curve"—the tendency of firms to reduce costs as they gain experience in manufacturing a new product (Argote and Epple 1990)—provides systematic quantitative support for this proposition.

Third, the models in this chapter, by focusing on the quality of policies rather than on agents' subjective preferences, implicitly presume risk neutrality. But if decision makers are risk averse, they may prefer incremental search simply because it reduces uncertainty by generating less dispersed distributions of options. However, sweeping assertions about the effect of risk aversion may be unfounded. Though it is tautologically true that a risk-averse agent prefers, for example, a sure thing to a lottery with the same expected value, this is not the choice that confronts decision makers in status quo–based policy processes. In these contexts, one can reject the new proposal and keep the status quo policy in force. By providing a lower bound on what the decision makers will get, the status quo truncates the distribution of new options by chopping off part of the "bad tail." This differs significantly from the choice between a sure thing and a lottery: if one chooses the latter, there is no lower bound other than the lottery's minimal value, leaving one fully exposed to downside risk. Thus, introducing risk aversion would not yield a conclusive answer favoring more incremental search. Risk-averse decision makers in status quo–based policy processes would have to trade off reducing a bad (more dispersion) versus increasing a good (upside risk).

Finally, in translating an informal theory into a formal model one always runs the risk of failing to represent the former correctly. Clearly, one property of Lindblom's informal theory—the trade-off between many frequent small steps and few infrequent big ones (Lindblom 1959, p. 86)—is not captured by the models here. This is a plausible trade-off: even an institution accustomed to parallel processing, such as the U.S. Congress, cannot pass many major bills in the same session. Major legislation requires a scarce resource—the attention of leaders. One could partly capture this by assuming that incremental search could occur every period but nonincremental search only every $s$ periods. Obviously, if $s$ is sufficiently large, then incrementalism is better.[25]

In any case, the models presented here are only a first step in clarifying the logic and claims of incrementalism. By the nature of formal models, they make clear which aspects of the complex strategy of muddling through have been represented and which omitted. This should make further steps (large or small) easier.

## CHAPTER 5

# The Perfect Is the Enemy of the Best

*Adaptive versus Optimal Organizational Reliability*

JONATHAN BENDOR AND SUNIL KUMAR

*The perfect is the enemy of the good.*

Voltaire

## INTRODUCTION

The study of organizational reliability is characterized by two possibly inconsistent themes.

(1) Scholars working in this area (Landau 1969; Bendor 1985) usually argue that individual decision makers are imperfectly rational and hence do not optimize when confronted with difficult problems.

(2) Nevertheless, some scholars also analyze the design of optimally reliable organizational systems (Heimann 1993, pp. 427–28, 434; Heimann 1997, p. 83; Ting 2003).

There appears, at least at first glance, to be some tension between these two themes. After all, designing high-reliability agencies is a hard problem: on this, both optimists (LaPorte and Consolini 1991) and pessimists (Perrow 1984; Sagan 1993) agree. So claim 1 implies that agencies will rarely be optimally reliable.

A straightforward way to resolve the tension is to insist on a clear distinction between descriptive and normative theories. Claim 1 is a descriptive assumption. Of course it has normative implications: as Landau argued, since individuals are imperfectly rational, if one wants to improve organizational reliability one should take structural solutions (such as redundancy) seriously. But fundamentally one should regard claim 1 as a building block of a descriptive theory of organizational reliability. This should be contrasted with normative theories of reliability, which naturally use the criterion of optimality.[1]

Once we have clearly separated the two types of theories, we can then go back and use them in tandem by analyzing whether agencies that are covered by a descriptive theory of reliability also satisfy certain normative criteria. In short, under what conditions would we predict that real agencies will be optimally reliable? Answering this question requires constructing both descriptive and normative theories—and it also requires that we remember that they are distinct formulations.

The contribution of this chapter is twofold. First, it establishes the validity of a general normative prescription—a property that all adaptive organizations must satisfy if they are to be optimally reliable. Second, it offers a descriptive analysis of a ubiquitous class of adaptive agencies that do *not* satisfy this normative prescription. Such agencies do not become optimally reliable even in the long run.

Our descriptive model will be consistent with claim 1: it presumes no fully rational "system designer" who optimizes trade-offs between different types of errors. Instead, the agency is adaptively rational (Cyert and March 1963): it adjusts its behavior when it confronts reliability problems. These adjustments are prima facie sensible: they involve heuristics that appear reasonable for the task at hand. One important rule of thumb represented by our descriptive theory has a *putting-out-the-fire* quality, in the spirit of Cyert and March's account of problemistic seach. Consider, for example, a central choice problem of the Food and Drug Agency (FDA): whether to approve or reject a drug. (We shall use this as a running example throughout the chapter.) Suppose that the agency has recently approved a drug that turns out to produce lethal side effects unacceptably often. This causes a public outcry and a crisis for the agency. One putting-out-the-fire response by the FDA would be to build in another (redundant) check point or hurdle that similar drugs would have to satisfy in the future. However, if a drug that was stalled in the agency dramatically proved its worth in another country, the organization might expedite the process (Heimann 1997, pp. 65–66). Thus, the agency would have ways of responding to both type I errors (accepting bad drugs) and type II mistakes (rejecting good ones).

Adaptation by putting out the fire is sensible behavior: if a problem occurs, try to ameliorate it. And we shall see that the heuristic does indeed have certain normatively desirable properties (proposition 1): it guides decision makers in the right directions when problems become more (or less) common. However, as we shall see in part (ii) of theorem 1, putting out the fire also lacks other, equally desirable properties. In particular, we shall see that organizational adaptation by putting out fires and related heuristics leads directly to what Heimann has called "cycles of failure":

> Ultimately, we see that a cycle between greater type I and type II reliability occurs within an agency. The agency begins with concern for type I reliability but shifts to type II reliability in due time. This shift does not last, however. Given limited resources, this shift toward type II reliability comes at the expense of type I reliability. Eventually, that means that a major accident or near-failure will occur and refocus public attention on the potential hazards of the technology. The political dictum of "be seen doing no harm" then prevails and pushes the agency to seek greater reliability against type I errors. The pressure for cost-effectiveness, suppressed for a while as a result of the accident or near-failure, re-emerges as the memory of the incident dims. The oscillation between type I and type II reliability continues as the agency now renews efforts to reduce the likelihood of a type II failure. (1997, p. 169)

Thus, it will turn out, as Heimann's verbal formulation suggests, that *relentless error correction*—the agency first putting out one fire (a type I failure) and then putting out another (a type II failure)—*is suboptimal*: it prevents the organization from finding, even in the long run, the optimal combination of the two types of reliability. Hence, as this chapter's first result (theorem 1) will show, it is possible for there to be too much adaptation.

In contrast, note that part (ii) of theorem 1 implies that any stationary rule must be suboptimal for the class of environments described by the hypotheses of the theorem. The reason is that any such rule is caught between Scylla and Charybdis: it is either perpetually restless, hence unable to settle down on optimal behavior, or always inertial, whence it is insufficiently exploratory and so may not discover what is optimal in the first place.[2]

This chapter is organized as follows. The following section presents a quite general stochastic model of an adaptively reliable organization. This model yields the chapter's first results (contained in theorem 1), which compare the behavior of adaptively and optimally reliable agencies in a stationary environment. The third section studies how the adaptive agency adjusts to changes in its environment. The answers to this difficult question (propositions 1–4) are generated by imposing a Markovian structure on the agency's adaptation. Thus, the model in this section is a special case of the one presented in the second section.[3] Extensions and conclusions are offered in the chapter's fourth and final section.

## A GENERAL MODEL OF AN ADAPTIVELY RELIABLE AGENCY

Consider an agency that approves or disapproves a series of proposals, one per period. These proposals may be internally generated, such as plans for dams in the Army Corps of Engineers, or externally generated, such as applications for drug approval at the FDA. The agency has $n$ levels of decisional standards, $(s_1, \ldots, s_n)$, which it uses to screen proposals. We denote the probability of a type I error, given a standard of $s_i$, by $\Pr(I|s_i)$; this probability is also conditioned on the project being bad. $\Pr(II|s_i)$ is the analogous conditional probability of a type II error, given a good project.

We say that a standard is *perfect* if, when the agency uses it, neither type I nor type II errors are ever made: that is, a standard $s_i$ is perfect if and only if $\Pr(I|s_i) = \Pr(II|s_i) = 0$. Otherwise we call the standard *imperfect*. (Note that it is possible for the probability of one of the two types of error to be zero under an imperfect standard.) Thus, in this model, an agency that is highly reliable with respect to type I errors is simply one that uses standards that rarely produce such mistakes. High reliability regarding type II errors has an analogous meaning. *Perfect* reliability is, of course, usually unattainable; typically we expect different standards to trade off one reliability for the other.

As a simple example of different decisional standards, suppose that the final decision on a proposal is made by a committee with $n$ members. Approval is given if at least $k$ people on the committee vote for the proposal.[4] Presumably the stringency of the process—the likelihood that a proposal will be rejected—is increasing in $k$. However, while this monotonicity assumption will be used in the more specialized model in the third section, it is not required for the more general model of this section. All that theorem 1, below, requires is that no $k$-out-of-$n$ system be perfect.

Moreover, the above voting scheme is only one example of how the agency can impose different standards on proposals. There are many other ways. For example, it could impose burden-of-proof rules, such as those used in legal institutions: the project must be worthwhile beyond a reasonable doubt rather than have only a preponderance of evidence in its favor.

For simplicity we assume that there are two types of projects, good and bad. (Extending to more than two types is straightforward, but all of our qualitative insights can be obtained with just two types.) A good project is one that an agency staffed by fully informed and completely rational decision makers would accept. A bad project is one that such an agency would reject. A project's quality is independent of all other actions and outcomes in the system. We further assume, for the agency in a stationary environment, that there is a constant probability that the project is good, $p_g \in (0, 1)$. Naturally, the chance that the project is bad, $p_b$, equals $1 - p_g$. (In the third section we examine the effects of changes in these probabilities.)

We assume that both types of errors impose costs on the agency. However, we do not need to represent how these costs explicitly enter into the agency's utility function. Instead, we assume that there is an expected utility associated with any standard, and a fully rational agency would maximize this utility function. (No specific assumptions about the agency's attitudes toward risk are required.) However, we follow the convention in reliability theory by focusing instead on the so-called *loss function*, which is strictly increasing in the costs and probabilities of mistakes. Because the loss function is essentially the inverse of a utility function, optimization entails trying to minimize this function. We denote the loss function by $L(\cdot)$.[5]

Note that this benchmark of a fully rational agency is *not* equivalent to the much stronger claim that the agency is maximizing social welfare. Instead, it involves only the weaker assumption that there is *some* utility function which the bureau maximizes. Thus there may be considerable slippage between the benchmark agency's preferences and social welfare. (For more on this issue, see the conclusion.)

Our adaptive agency does not know what the true probabilities are—it is precisely this uncertainty that mandates some form of adaptation—nor does it attempt to maximize or minimize any objective function.[6] Instead, it responds to errors, in a simple and rather intuitive way. Let $q^I_{i,j;t}$ denote the probability that, having discovered a type I error in period $t$, the agency changes the standard from level $i$ to level $j$; let $q^{II}_{i,j;t}$ be analogously defined for type II errors. The main assumption of our general model says that if the agency discovers that it has made a mistake, then there is some chance, bounded away from zero, that it will adopt a new standard in the next period. In short, *errors induce search* (for new standards).

(A1)   *There is an $\epsilon > 0$ such that for each $i$ and for all $t$, $1 - q^I_{i,i;t} \geq \epsilon$ and $1 - q^{II}_{i,i;t} \geq \epsilon$.*

One might think of (A1) as stating that the agency will persistently engage in trial-and-error behavior (unless, of course, it hits upon a perfect standard). Note that (A1) does not stipulate *how much* the agency adjusts. Observe also that (A1) allows for the possibility of inertia: the agency might not change its standard after having discovered an error. All that is ruled out is deterministic inertia, following negative feedback.

A well-known adaptive scheme that satisfies (A1) is the Bush-Mosteller rule. Suppose that the agency has two options, $a$ and $b$. Let $p_t(a)$ denote the probability that $a$ is used in period $t$. Suppose that the agency uses $a$ in $t$. Then the Bush-Mosteller rule adjusts the agency's propensity to use $a$ in the next period as follows. If using $a$ resulted in the correct decision, then $p_{t+1}(a) = p_t(a) + \alpha(1 - p_t(a))$, where $\alpha$, a parameter representing the speed of learning given success, is in $(0, 1)$. If using $a$ produced the wrong decision, then $p_{t+1}(a) = p_t(a) - \beta p_t(a)$, where $\beta$, the speed of learning in the face of failure, is also in $(0, 1)$. Analogous equations describe the agency's response to feedback if it has used option $b$. Note that the second equation, which captures how negative feedback is handled under Bush-Mosteller, satisfies (A1).

Finally, note that (A1) says that the agency's adjustment is *problem driven*: it might change its standard if it confronts a problem (i.e., an error). It does not say what happens if the agency gets no information about the quality of the proposal whose fate was decided in $t$ or if the agency does get feedback and it is positive (i.e., it shows that the agency's prior choice was correct). No such assumption is required for the main result of this section, theorem 1. (We do, however, stipulate how the agency behaves under these eventualities—no feedback or positive feedback—in the more specific model in the third section of this chapter.) Here all we need to presume is (A1) and a complementary assumption that sometimes the agency will find out whether it made a mistake.

Let $d^a$ denote the exogenously fixed probability that the agency will discover or be told the quality of a proposal that it has accepted. Similarly, $d^r$ denotes the probability that the agency will discover the quality of a rejected project. (It is plausible to assume that $d^a > d^r$, but we do not need that assumption.) Both probabilities are in $(0, 1)$ and are independent of the agency's choice behavior and of all other parameters of the model. Since these probabilities exceed zero, and since $p_b$ is in $(0, 1)$, the agency will with positive probability discover both type I and type II errors.

Note that we have *not* assumed that the agency's adaptation is Markovian: (A1) allows for $q^I_{i,j;t}$ and $q^{II}_{i,j;t}$ to depend on events that occurred in the agency's distant past.

We now turn our attention to evaluating the performance of an agency that behaves in the above manner. We have intentionally chosen a weak normative criterion (below) for two reasons: first, it is a necessary condition for optimality in any reasonable sense of that word; second, we will show that satisficing-type rules do not satisfy it. Failure to satisfy a normative criterion is more interesting if the criterion is weak.

DEFINITION 1.   *The agency is* potentially optimal *if there is some finite date T such that with positive probability* $S_t$ *is optimal for all* $t > T$.

Hence, if an agency is not even potentially optimal, then it will *with certainty* use a suboptimal standard infinitely many times. Thus, a necessary condition for potential optimization is that the bureau avoid being permanently trapped (with probability 1) on a suboptimal action. (We say that the agency is permanently trapped on a suboptimal action $s_i$ if there is a finite date $T$ such that for all $t > T$, $S_t = s_i$.) Thus, the bureau must be sufficiently inclined to search for a new standard if the current one is, in some sense, unsatisfactory. The benign aspect of (A1) is that it achieves this necessary condition, as part (i) of theorem 1 establishes. However, this propensity of (A1)-type rules to explore new options, following an unsatisfactory experience, is a two-edged sword. Its downside is conveyed by part (ii) of the theorem.

THEOREM 1.   *Suppose that no standard is perfect. Then with probability 1, no adaptive rule that satisfies (A1) can get the agency permanently trapped on any one action. Hence, with probability 1, the following conclusions hold.*

(i)   *No adaptive rule satisfying (A1) can get the agency permanently trapped on any one suboptimal action.*

(ii)   *If there is a uniquely optimal standard, then any adaptive rule that satisfies (A1) will use a suboptimal standard infinitely often. That is, no such rule is potentially optimal.*

Part (i) holds because any adaptive rule that falls into the class identified by (A1) is too restless to allow the agency to be stuck forever with an action that isn't optimal. Any such action must be imperfect; hence, it produces either type I or type II errors or both. And once an error is discovered, an agency using an (A1)-type rule will, with a probability bounded away from zero, search for a new standard. The proof then shows that the event of being permanently trapped on a suboptimal standard occurs with probability zero.[7]

However, as we see in part (ii), this restlessness exacts a price. It does not in general resolve the tension between exploration and persistence optimally because (A1) is insufficiently discriminating. Though no (A1)-type rule will be permanently satisfied with any particular *sub*optimal standard, no such rule will be permanently satisfied with *any* standard that is imperfect—not even with an optimal-yet-imperfect standard. Hence, the following result follows immediately from part (ii).

COROLLARY I.    *Any rule that is potentially optimal must violate (A1).*

Logically speaking, corollary I simply restates part (ii) of the theorem. But because the corollary highlights (A1)'s role, it is a useful way to state the necessary condition that constitutes our general normative prescription.

The bad news of part (ii) is that it requires a set of assumptions that are only a bit more restrictive than those used to generate the good news of part (i): the only additional premise is that there is a unique optimal standard. Assuming uniqueness is quite mild. Note that the finiteness of S automatically ensures the existence of *some* standard that minimizes the loss function. Hence, all that the new condition does is eliminate the possibility that one standard is exactly as good as another, which is a knife-edge condition.[8]

Theorem I's bad news continues to hold if the agency's feedback about proposal quality can be incorrect. Indeed, assuming, as we do throughout this chapter, that if the agency gets feedback, then the information is accurate makes it *harder* to prove part (ii). Obviously, incorrect feedback makes it more difficult for an adaptive agency to converge to the optimal standard.

It is often argued, sometimes with much hand waving, that if an adaptive decision maker repeatedly faces the same choice problem, then he or she must eventually converge to optimal behavior. The significance of part (ii) of theorem I, then, lies partly in showing that it ain't necessarily so. The agency in the present chapter uses a heuristic from a large and rather plausible class of rules, but even in a stationary environment, where it faces the same choice problem over and over, it does not converge to optimal conduct even in the long run.

What is going wrong?

In a stationary environment the completely rational and fully informed decision maker can pick the optimal standard and stick with it. That this behavior is optimal follows immediately from the assumption that its task environment is stationary: since the agency faces exactly the same problem in $t + 1$ as it did in $t$, whatever standard was optimal in $t$ must remain optimal in $t + 1$.

This sounds innocuous enough. But consider its implications for a real agency—say, the FDA. Suppose the FDA approves a drug that turns out to have deadly side effects. These occur often enough to outweigh the drug's benign effects. Thus, the FDA's approval was a mistake, and the drug should be pulled from the market. But what are the implications for the FDA's regulatory standards? The correct answer is that this "fiasco" (as the media might call it) does not necessarily have *any* implications for the agency's decision-making standards. Why? Because we know that in general even an optimizing bureau is going to make errors of both types, so the appearance of a type I error does not necessarily disconfirm the hypotheses that (1) the agency's environment—in particular, the quality distribution of new drugs—is basically stationary, and (2) the organization's level of stringency is about right for this environment. Of course, either hypothesis *might* be wrong, but the appearance of a type I error does not settle the matter.

To put it more sharply and, perhaps, more counterintuitively, if the FDA behaved completely rationally, it would make mistakes at an optimal rate and, given the technology of drug assessment, *this optimal rate exceeds zero.*[9] Indeed, in an operations research model of this type of problem, it is common to specify the optimal decision rule in terms of the optimal (nonzero) rate of error.

Of course, an agency's decision technology can always be improved, but in the short run, it is a constraint like any other resource. Furthermore, the notion that technological improvements in drug assessment will produce error-free regulation is a fantasy. There will always be mistakes, if not of one type then of the other—and probably both.

This is a hard notion to accept. Americans in particular are optimists about the possibility of technological perfection. *But this optimism imposes unreasonably high expectations on the agency.* Consider Paul Quirk's description of the tasks facing the FDA: "In order to understand the behavior of the FDA, one must first appreciate the difficulty of the choices it must make. These choices often have the character of a dilemma: no matter how good the analysis is, or how upright the intentions may be, none of the alternatives are even moderately satisfactory" (1980, p. 203). Yet from the Olympian perspective of rational decision making, the FDA does not face a dilemma. It simply must balance the expected costs of one type of error against those of another type. Whichever regulatory standard does so most effectively is the standard of choice.

Hence, Quirk's description is puzzling, if we assume that he was using a rational choice theory. But if we attend closely to his words, we can understand what he meant and what decision theory he had in mind. Note especially the conclusion: "none of the alternatives are even moderately satisfactory." This is clearly an aspiration-based theory: as in Simon's satisficing model, Quirk entertains certain evaluative thresholds, probably for both types of errors. Regulatory outcomes that exceed these thresholds are considered satisfactory; those that fall short are not. The so-called dilemma arises because if the FDA tries to achieve one threshold, it falls short of the other. Indeed, if Quirk (or the FDA's political environment) holds very ambitious thresholds, then the agency may be incapable of satisfying either aspiration.[10] So nothing the FDA does is "even moderately satisfactory."

Unfortunately, while this aspiration-based viewpoint is quite sensible, overly ambitious aspirations can have perverse effects, as we saw in chapter 3. In particular, they can make it impossible to sustain optimal behavior, by a variant of Voltaire's "the perfect is the enemy of the good." Theorem 1's part (ii) tells us that demanding perfection typically prevents us from securing what is optimal.

Although our model does not explicitly represent aspirations, they are implicit in our adaptive agency's never-ending restlessness in response to negative feedback. This behavior is tantamount to the agency's believing (with positive probability) that *any error is unacceptable*. The discovery of a mistake implies that something is wrong with the status quo; it should be changed. As physiologists might say, the agency is too irritable: it never stops responding to negative stimuli.[11] There is, in short, something to be said for inertia.

Part (ii) of theorem 1 establishes that a large and sensible class of adaptive rules, which includes putting-out-the-fire heuristics, does not guide the agency to optimal reliability. However, neither part (ii) nor part (i) shows that there exists *any* adaptive rule that would do the job. (Recall that part [i] showed only that [A1]-type rules pass the test of a *necessary* condition.) Perhaps none is known? If so, then theorem 1 by itself would be seriously incomplete, as it does not imply that the heuristics satisfying (A1) are inferior to any other (known) rule.

Fortunately, however, we do not labor under such ignorance: there *are* adaptive rules that are known to be potentially optimal. It is worthwhile examining one of these, in order to compare its properties with adaptive rules that involve persistent adjustment. This comparison will give us insight as to why heuristics that satisfy (A1) are not optimal even in the long run and also what kinds of properties suffice for convergence to optimal reliability.

Consider the stark case in which there are only two standards, low and high, where given the low standard the agency accepts all proposals and under the high one it rejects all of them. Assume that the agency discovers the quality of the project after it makes its choice. Hence choosing level $s_1$ generates a loss $l(s_1)$ with probability $p_b$ and zero otherwise. Similarly, choosing level $s_2$ generates a loss of $l(s_2)$ with probability $p_g$ and zero otherwise. If $p_b$ (and hence $p_g$) were known, the rational decision maker would choose $s_1$ if $l(s_1) \cdot p_b < l(s_2) \cdot p_g$ and $s_2$ otherwise. As before, however, we consider the case in which $p_b$ is unknown. Then adaptation is called for, since one must try level $s_1$ in order to learn about $p_b$. This problem is mathematically equivalent to the Bernoulli two-armed bandit problem.[12] While the details of the analysis of this problem need not be repeated here (see Rothschild 1974), the key conclusions are as follows.

First, if $l(s_1)p_b \neq l(s_2)p_g$, then with positive probability the optimal adaptive scheme settles down on the optimal standard after finitely many periods and stays at that level thereafter. In other words, the optimal adaptive scheme is potentially optimal, in the sense of definition 1. Second, the optimal scheme relies on a memory-intensive performance index of the $s_i$: the index uses *all* past observations (suitably discounted), not just the most recent one. The index (usually called the Gittins index, to honor the idea's inventor [see Whittle 1982]) tells the agent what action should be taken in every period. It is vital to note that because of the index's memory-intensiveness, this adaptive scheme becomes *less and less restless as time passes*, eventually settling down on the optimal standard with positive probability.[13]

Rather than describing the Gittins index rule in more detail, we find it more instructive to outline its behavior in this two-armed bandit problem. Initially, the Gittins rule explores both actions; then it settles down on one of them. Typically, it selects the optimal arm. However, it may settle on the suboptimal action—for example, after the bad arm has generated a freakishly long run of good outcomes. But the agent is tricked into selecting the suboptimal action exactly when that inferior arm paid off handsomely during the exploration phase. Hence, from the perspective of (say) ex ante minimization of the expected sum of discounted losses, this settling on the bad arm does not make the rule suboptimal.

The Gittins rule is quite simple, even unsophisticated in some respects. It does, however, have a very long memory, and *it is stoic in the face of failures*, in contrast to the never-ending adjustment of putting out the fires and other (A1)-type rules.[14]

This example is in a simple setting. There are two reasons for this. First, we use this example in the spirit of an existence proof, to establish that the notion of "potentially optimal" is not vacuous: it is possible to construct such schemes, at least for certain problems. Second, there is a substantive problem in more general settings where $d^a$ and $d^r$ are less than 1. In those contexts, one can only estimate the products $p_b d^a$ and $p_g d^r$, not the products' individual components. Hence, it may not be possible to construct any scheme that is potentially optimal. To this, consider the extreme case in which the agency never finds out the quality of rejected proposals: $d^r = 0$. In this case only type I errors are observed, and hence any adaptive scheme will settle on the higher standard $s_2$ (with probability $1$) since it never observes a type II error. Clearly, this may be suboptimal: type II errors occur; they just go unobserved.

In the other case, when the loss function $L(s_i)$ depends on $p_b$, $d^a$ and $d^r$ only through the products $p_b d^a$ and $(1 - p_b)d^r$, optimal adaptive rules exist, as illustrated by the above special case. Note that this case corresponds to two quite different circumstances: either perfect feedback ($d^a = d^r = 1$, as in the preceding simple analysis) or feedback is imperfect but the decision maker's utility function depends only on observed outcomes, not actual ones. The latter might approximately describe agencies that do not care about much more than appearances.

A MARKOV MODEL OF ADAPTIVE RELIABILITY

Theorem $1$'s part (ii) tells us, among other things, that persistent putting-out-the-fire behavior does not produce optimal reliability in a stationary environment. How does it perform if the environment's basic parameters change? To answer this question, it is helpful to impose more structure on the process. So we now assume that the agency's adjustment is Markovian: both $q^I_{i,j;t}$ and $q^{II}_{i,j;t}$ depend only on the state $i$ and what happened in period $t$. They may not depend on what happened in the bureau's history prior to $t$. Further, we assume that this is a Markov process with stationary transition probabilities, so we write $q^I_{i,j;t} = q^I_{i,j}$ and $q^{II}_{i,j;t} = q^{II}_{i,j}$. These assumptions are summarized by (A2).

(A2)   *The agency's adaptation is governed by a stationary Markov process over finitely many states, $(s_1, \ldots, s_n)$. The process is represented by the probability transition matrix P depicted in figure 6.*

To complete the description of the Markov process, we add three important substantive assumptions. The first concerns the technology of decision making.

(A3)   *The probability that a proposal will be accepted is strictly decreasing in $s_i$.*

(A3) imposes a well-defined ordering on the standards. (Hence, it is now appropriate to call them *stringency levels*.) The toughest standard is the highest one, $s_n$; the weakest one is $s_1$. A natural example of this is the $k$-out-of-$n$ voting system, where $k$ denotes the number required for approval. If the decision makers vote sincerely and they err independently, then clearly the probability of rejecting proposals is increasing in $k$.[15] In terms

The System in Period $t+1$

| | 1 | $\ldots$ | $i-1$ | $\ldots$ | $i$ | $i+1$ | $\ldots$ | $n$ |
|---|---|---|---|---|---|---|---|---|
| | $\ldots$ | $\ldots$ | $\ldots$ | | $\ldots$ | $\ldots$ | $\ldots$ | $\ldots$ |
| $i$ | $p_g\Pr(II\|S_i)d^r q_{i,1}^{ll}$ | $\ldots$ | $p_g\Pr(II\|S_i)d^r q_{i,i-1}^{ll}$ | | $1 - p_g\Pr(II\|S_i)d^r(1-q_{i,i}^{ll}) -$ $p_b\Pr(I\|S_i)d^a(1-q_{i,1}^{l})$ | $p_b\Pr(I\|S_i)d^a q_{i,i+1}^{l}$ | $\ldots$ | $p_b\Pr(I\|S_i)d^a q_{i,n}^{l}$ |
| | $\ldots$ | $\ldots$ | $\ldots$ | | $\ldots$ | $\ldots$ | $\ldots$ | $\ldots$ |

The System in Period $t$

Figure 6. The probability transition matrix of the Markov process.

of redundancy theory (Landau 1969), if $k = n$ then the system is highly redundant with respect to type I errors: the organization will mistakenly accept a bad proposal only if all the decision makers make that mistake.[16]

The other two substantive assumptions concern the nature of adjustment. (A4) says that the agency satisfices (Simon 1955): change is *failure driven*. In the view of an agency that adjusts according to (A4), "if you don't know whether it's broke, don't fix it."

(A4)   *The agency increases its standard in t only if it discovers that it made a type I error in t − 1 and decreases its standard only if it discovers that it made a type II error in t − 1.*

In addition to incorporating satisficing, (A4) has a directional feature related to the putting-out-fires heuristic. It implies that the agency will not decrease its stringency level if it finds out that it has just accepted a bad proposal, nor will it become more stringent if it discovers that it has just rejected a good one. In view of (A3), such adjustments seem perverse—at least from the common-sensical perspective of putting out fires—and so are ruled out.

The next assumption says what *can* happen if an error is found.

(A5)   *For all $i < n$, $q^I_{i,i+1} > 0$; for all $i > 1$, $q^{II}_{i,i-1} > 0$.*

(A5) contains a second directional hypothesis: if the agency discovers that it accepted a bad project, then it might tighten up its standard, by going from the status quo of $s_i$ to $s_{i+1}$. Similarly, if it finds out that it rejected a good proposal, then it might relax its standard, by going from $s_i$ to $s_{i-1}$. This directional feature is simply the other (positive) side of putting out fires. It too is partly motivated by the technological assumption of (A3): moving to higher standards does make type I errors less likely, and (A5) says that the agency behaves as if it knows that technological fact and adjusts accordingly.

(A5) incorporates a second premise: putting out the fire might involve the smallest possible tightening or loosening of the agency's standard. This assumption is clearly consistent with notions of incremental decision making (Lindblom 1959); as such, it is squarely within the research program of bounded rationality, as Lindblom made clear. Note, however, that (A5) does not require that adjustment be local; it stipulates only that it is *possible*. Hence, it comports with theoretical reformulations of incrementalism (e.g., Padgett 1980; Bendor 1995) that allow for nonlocal adaptations.

Given that the transition matrix $P$ satisfies (A3) and (A5), it is easy to show that all states communicate. Hence a basic theory of finite

Markov chains tells us that the process has a unique invariant probability distribution—that is, a vector $\Pi = (\pi(s_1), \ldots, \pi(s_n))$ that is the unique solution to the steady-state equation $\Pi P = \Pi$. Further, because both good and bad proposals occur with positive probability and because both $d^a$ and $d^r$ are less than 1, every cell on $P$'s main diagonal is strictly positive, so the Markov chain is aperiodic. Because the process has a unique invariant distribution and is aperiodic, it must be *ergodic*: it must converge to that (unique) probabilistic equilibrium from any initial vector of probabilities over the states (see for example, Kemeny and Snell 1960). This fact will be very useful to us in the following results.

## Responses to a Change in the Quality of Projects

First we consider how an agency that puts out fires will respond to a change in the underlying distribution of the quality of projects—that is, to changes in $p_g$. Proposition 1 shows that this class of heuristics does yield adaptively rational outcomes: the bureau responds to such changes in a manner that is qualitatively consistent with optimization.

Instead of comparing a single agency over time, it we find convenient to compare two agencies, X and Y, that are in different environments. For agency X the probability that a proposal before it is good is $p_g(X)$, whereas for agency Y it is $p_g(Y)$. A convenient abuse of notation is that $X_t$, a random variable, denotes agency's stringency level in $t$, and analogously for $Y_t$.

Suppose, then, that $p_g(X) > p_g(Y)$: what should happen? The direct effect is clear: since agency Y encounters more bad projects than X does, if in date $t$ the agencies are using the same standard, then today Y is more likely to make a type I error than X is, which implies that in $t + 1$ agency Y will be more likely to tighten its standards than will X. But this creates an indirect effect as well: this greater tightening of standards makes Y *less* likely than X to accept a bad project in $t + 1$, by the monotonicity assumption of (A3). The very fact that Y faces a worse pool of alternatives induces that bureau to adapt by becoming tougher than X, which in turn reduces the chance that it will make a type I error in the first place.

The net result of the direct effect of an inferior pool of options and the indirect effect of the agency's adaptation is not obvious. However, the following assumption, (A6), tilts matters decisively.

To state this assumption concisely, we again use the idea of stochastic dominance first encountered in chapter 4. Here we say that one probability distribution over the $n$ stringency levels, say $Y_t$, stochastically dominates another distribution, $X_t$, if (1) $\Pr(Y_t > i) \geq \Pr(X_t > i)$ for all

$i = 1, \ldots, n - 1$, and (2) at least one of the inequalities in (1) is strict. (Recall that if $Y_t$ stochastically dominates $X_t$, then $E[Y_t] > E[X_t]$.) Thus, because distribution $Y_t$ is shifted to the right of distribution $X_t$ in at least one place, in a clear probabilistic sense $Y_t$ is more stringent than is $X_t$. If it is known only that property (1) holds, then we say that $Y_t$ weakly stochastically dominates $X_t$.

We use the idea of weak stochastic dominance as an assumption about the conditional distributions over the stringency levels identified by the rows in the transition matrix $P$. (The stronger version of stochastic dominance is derived, not assumed: it emerges as part of the conclusions of propositions 1–3.) In the following assumption, let $p_{i,j}$ denote the probability of transitioning from state $i$ to state $j$ in $P$.

(A6) *In matrix $P$ each row weakly stochastically dominates its predecessor: for all $i > 1$, $\sum_{j=k}^{m} p_{i,j} \geq \sum_{j=k}^{m} p_{i-1,j}$ for $k = 2, \ldots, n$.*

To get a feel for what (A6) means, consider a simple special kind of agency adaptation: purely incremental adjustment. If an agency discovers that it has made a type I error, then with probability $p > 0$ it increases its stringency by one level, from $s_i$ to $s_{i+1}$ (if the current standard isn't already maximally stringent). If it discovers that it has made a type II error, then with probability $p$ it reduces its standard from $s_i$ to $s_{i-1}$ (if the current standard isn't already as lax as possible). In either case, with probability $1 - p$ it stays at $s_i$. It is easy to see (figure 7) that the probability transition matrix of this purely incremental adjustment satisfies (A6)—and more: if we compare row $i$ to row $i - 1$, we see that tomorrow's probability distribution over stringency levels, if we are at level $i$ today, is for certain state values pushed to the right, compared to tomorrow's probability distribution given that today we are only at level $i - 1$. Thus, row $i$ stochastically dominates row $i - 1$; it doesn't merely weakly dominate it.[17]

We now deploy our specialized model, (A2)–(A6), in the next result. Proposition 1 will include one result about the limiting behavior of agencies (part [i]) and also a result about the *paths* of the agencies, before they reach their steady states (part [ii]). The latter is probably of greater empirical relevance.

PROPOSITION 1.    *Suppose (A2)–(A6) hold. Compare two processes, $X_t$ and $Y_t$, whose transition matrices are identical except that $p_b(Y) > p_b(X)$.*

(i)   *In the limit $Y_t$ stochastically dominates $X_t$.*

(ii)  *If we also assume that $X_0$ and $Y_0$ are identically distributed, then $Y_t$ stochastically dominates $X_t$ for all $t > 0$.*

The System in Period $t + 1$

| | 1 | ... | $i-2$ | $i-1$ | $i$ | $i+1$ | $i+2$ | ... | $n$ |
|---|---|---|---|---|---|---|---|---|---|
| ... | ... | ... | ... | ... | ... | ... | ... | ... | ... |
| $i-1$ | 0 | ... | $p_g\Pr(II|S_{i-1})d^r q^{II}_{i-1,i-2}$ | $1 - p_g\Pr(II|S_{i-1})d^r q^{II}_{i-1,i-2} - p_b\Pr(I|S_{i-1})d^a q^l_{i-1,i}$ | $p_b\Pr(I|S_{i-1})d^a q^l_{i-1,i}$ | 0 | 0 | ... | 0 |
| $i$ | 0 | ... | 0 | $pg\Pr(II|S_i)d^r q^{II}_{i,i-1}$ | $1 - p_g\Pr(II|S_i)d^r q^{II}_{i-i} - p_b\Pr(I|S_i)d^a q^l_{i,i+1}$ | $p_b\Pr(I|S_i)d^a q^l_{i,i+1}$ | 0 | ... | 0 |
| $i+1$ | 0 | ... | 0 | 0 | $p_g\Pr(II|S_{i+1})d^r q^{II}_{i+1,i}$ | $1 - p_g\Pr(II|S_{i+1})d^r q^{II}_{i+1,i} - p_b\Pr(I|S_{i+1})d^a q^l_{i+1,i+2}$ | $p_b\Pr(I|S_{i+1})d^a q^l_{i+1,i+2}$ | ... | 0 |
| ... | ... | ... | ... | ... | ... | ... | ... | ... | ... |

The System in Period $t$

Figure 7. The purely incremental process.

The results reported by proposition 1 represent the same directional adjustment as an optimizing agency would make. That is, suppose such an agency set its standard to optimally trade off the expected costs of type I and type II errors. Then imagine that the probability of bad projects increased for exogenous reasons. If the agency does not alter its standard, then the expected costs of type I errors will rise while those of type II will fall. To bring them into closer balance once again, the fully rational agency will—if it can adjust its decisional standard to a sufficiently fine degree—tighten its standards.[18] Hence the fully and the imperfectly rational bureaus move in the same direction in response to this important change in their environments.

This has an interesting implication for the empirical study of bureaus. Perhaps the most common way to test rational choice theories of institutions is to use regression analysis to test the model's comparative static predictions. In comparative statics, one compares an agent's optimal behavior for different values of a given parameter, as in the above example of an exogenous increase in the frequency of bad projects leading to a higher optimal stringency level. In the most basic version of the ensuing regression analysis, the analyst tests to see whether the coefficient has the right sign: does the agent's choice variable move in the predicted direction in response to the parameter's change? But note that here the model of adaptive reliability predicts a shift in the same direction. Hence, empirical scholars, beware: a simple regression study—one that reports only the sign of the relevant coefficient—does not discriminate between the hypotheses of full rationality and the kind of imperfect rationality embodied in the putting-out-fires heuristic. Thus, what is good normative news is bad methodological news.

*Responses to External Pressures to Approve Projects*

A regulatory agency such as the FDA often must deal with interest groups that want changes in the agency's decision-making process. Pharmaceutical companies, for example, typically want easier drug-approval processes.[19] Thus, they may lobby Congress to pass legislation that lowers the burden of proof that new drugs must meet. Suppose they succeed in getting such legislation enacted. How will adaptive agencies respond? The next result provides the answer.

PROPOSITION 2 (LEANING AGAINST THE WIND). *Suppose that (A2)–(A6) hold. Assume also that for every standard the probability that agency Y accepts proposals is higher than the probability that agency X accepts them.*

   *(i)   In the limit $Y_t$ stochastically dominates $X_t$.*

  *(ii)   If we also assume that $X_0$ and $Y_0$ are identically distributed, then $Y_t$ stochastically dominates $X_t$ for all $t > 0$.*

Thus, agencies that operate in political environments that push for project approval adapt by becoming tougher.[20] The intuition for this is straightforward. If the burden of proof confronting new options falls, then both bad and good proposals will be approved more often. Hence, the agency is more likely to make type I errors and less likely to make type II errors. Given the adjustment heuristic, increases in the first type of mistake make the agency more prone to tighten its stringency level, and decreases in the second type of mistake make the bureau less inclined to relax its regulatory stringency.

Next we consider how the agency responds to technological or scientific improvements in its ability to recognize when it has rejected good proposals (or, by implication, how it responds to degradation in this ability). Again, consider drug regulation. It is widely believed (e.g., Heimann 1997, p. 65; Quirk 1980, pp. 217–18; Lohmann and Hopenhayn 1998, p. 239) that it is easier for the FDA to find out that it has approved a harmful drug than to discover that it has rejected a benign one. The reason for the difference is obvious. An approved drug, harmful or otherwise, will be used by thousands or millions of Americans. Thus, there are many opportunities for discovering that a drug has, for example, deadly side effects. In contrast, good drugs rejected by the FDA are used less often in the United States, so getting field data about their efficacy or side effects is more difficult. However, the FDA is not the only regulator of pharmaceuticals: many other industrialized countries have similar agencies and, because their decision-making patterns differ, a drug blocked in one country may be approved in another. Suppose, then, that there are improvements in the exchange of information among agencies in different countries that regulate pharmaceuticals.[21] Then the probability that the FDA discovers that it has made a type II error will rise. The effect of this change is reported in the next result. (Since $d^r$ denotes the probability that an agency discovers the quality of a rejected proposal, $d^r(X)$ is the chance that agency X makes that discovery.) As usual, we compare two agencies adapting in different parametric environments.

PROPOSITION 3.   *Suppose that (A2)–(A6) hold, and also that $d^r(X) > d^r(Y)$.*

(i)   In the limit $Y_t$ stochastically dominates $X_t$.

(ii)  If we additionally assume that $X_0$ and $Y_0$ are identically distributed, then $Y_t$ stochastically dominates $X_t$ for all $t > 0$.

The FDA's experience with anti-AIDs drugs partially corroborates this result.[22] As AIDs groups and the FDA became aware of information about potentially effective drugs used overseas, the agency put such pharmaceuticals on a "fast track" that made approval easier (Heimann 1997, pp. 154, 158).[23]

### Responses to Improvements in Error Prevention and Error Detection

Suppose that the technologies of error prevention and error detection improve. How will adaptive agencies respond to these changes? Answering this question turns out to be more difficult than answering the preceding three—we have been unable to derive a pathwise property—so our next result says only what happens in the limit.

It may appear natural to hypothesize that these technological improvements have important effects on how close the adaptive agency comes to optimal reliability, in the long run. If so, certain kinds of technological progress might make part (ii) of theorem 1 quantitatively unimportant. Suppose the following were true: as the probability of making either type of error falls, the probability that the agency chooses the optimal standard in the limit rises. Then although strictly speaking it would still be true that for any strictly positive error probabilities the adaptive agency would not converge to optimal choice, the steady-state difference between adaptively and fully rational decision making might be small.

This is an interesting conjecture. But it is false. The next result shows that in general lower rates of making errors, or higher rates of correcting them, do not imply a greater chance of finding the best stringency level in the long run. To establish this we specialize the Markov model by presuming that adjustment is always incremental (i.e., $q_{i,j}^I = q_{i,j}^{II} = 0$ if $j$ exceeds $i + 1$ or is less than $i - 1$).

PROPOSITION 4.   Suppose (A2)–(A6) hold and the agency's adjustment is always incremental. Assume that agency Y makes both type of errors less often than X does; for every stringency level, Y's rate is $\alpha$ times X's, where $0 < \alpha \leq 1$. Further, suppose that X detects errors less often: its rate is $\beta$ times Y's, where $0 < \beta \leq 1$. Despite these advantages of Y, the agencies' long-run propensities over standards are identical.

Since X and Y reach the same long-run propensities concerning *all* possible stringency levels, proposition 4 immediately implies that Y is no more likely to use the optimal standard in the limit than X is.[24] The intuition for the result is as follows. Suppose $\alpha = \frac{1}{2}$ and $\beta = 1$. Then, in every circumstance, Y is only one half as likely as X to approve bad proposals or reject good ones. True, this means that Y, having reached the optimal standard at a particular date, is less likely to leave it, since only discovered errors will induce it to change its rule, and X is more error prone and just as good at detecting its mistakes. But by the same token, *and to the same degree*, X is more likely to discard suboptimal standards, and so is more likely to hit upon the best one. Hence, since its rates of both entering and leaving the optimal state is twice Y's, its invariant propensity to be at the optimal standard must exactly equal Y's.

Similarly, suppose that Y and X make errors equally often but Y is better at error detection: it is more likely to discover its flawed choices. Now Y's advantage produces the mirror image of the preceding discussion: better error detection makes Y more likely to move into the optimal state but also more likely to move out of it.

This explanation, which is based on rates of leaving and entering states in a Markov process, may satisfy readers with a taste for such analyses. A more intuitive explanation of why (for example) better error detection doesn't help the agency stabilize at the optimal reliability level is as follows. Recall that the best approach in a stationary environment is to try to figure out what is the optimal standard and then, having found it, *to stick with it*. In contrast, if our adaptive agency always discovers its mistakes, then—given its adjustment heuristic—anytime it errs it may change its stringency level in the next period. While this is desirable if (in the above example) its stringency had been lax or tight, it is undesirable had it been using a middling (optimal) standard. Since it is inevitable that even the best stringency level will sometimes result in errors, a bureau that flawlessly corrects its mistakes must eventually spurn the optimal standard. Of course, ceaseless error detection also ensures that it must sometime move *toward* the optimal stringency level as well. But given its unreasonably high implicit aspiration level—no error is acceptable—it cannot stay there. Thus, tireless error detection is suboptimal.

Of course, making errors less often or detecting them more frequently is intrinsically desirable: proposition 4 does not gainsay that obvious truth. What it does tell us is that such improvements do not inevitably confer a second, indirect benefit: they do not help an agency find, even in the very long run, the right decisional standard.

CONCLUSIONS

Several of our assumptions are rather stark. For example, proposals in the real world come in greater variety than merely either good or bad: some might be disasters waiting to happen; a few could be genuine break-throughs. And the agency's response to errors would likely depend on the magnitude of the failure: for example, if the agency had approved a Thalidomide, one would expect it to respond more vigorously than if it had approved a drug that is deficient in minor ways.

We believe, however, that the thrust of our argument does not depend on the above stylized premises. In particular, we conjecture that theorem 1's part (ii) would remain valid if one extended our model to allow for more than two quality levels and to let the agency's response be a function of an error's gravity. We maintain this because the essential logic of theorem 1's second part—if no alternative is perfect, then never-ending adaptation produces suboptimal behavior—does not depend in any essential way on those premises. Problems could come in many sizes, and the agency could respond differently to big ones than to small ones. We hypothesize that as long as (A1) holds, the agency cannot stabilize at an optimal level of stringency. The main reason for the discrepancy between the real and the ideal would remain qualitatively the same: no matter what the agency does, there is always *some* chance that something will go wrong. And since the agency refuses to respond to fires by doing nothing, it ensures that it will adjust its standard indefinitely. Hence, it cannot stabilize at the optimal rule. The culprit once again is an implicit aspiration level that is unrealisti-cally ambitious: any error might be considered unacceptable. This produces relentless error correction in the pursuit of perfection. And when perfection is unattainable, pursuing it relentlessly is the enemy of the good. Indeed, Voltaire's remark can be strengthened: the perfect is the enemy of the best.

One might try to rescue the normative standing of putting-out-fire rules by arguing that (1) real agencies confront nonstationary environments and (2) in such contexts putting-out-fire rules are, in fact, optimal. Claim 2 is an interesting theoretical assertion. But presently it is, we believe, only a conjecture: we have never seen a proof showing that it is true. Regarding claim 1, it is important to understand what is at issue. To establish that putting-out-fire rules are optimal, it is not enough to argue that agencies such as the FDA face environments that are nonstationary in *some* respect. Obviously, the FDA encounters new types of drugs every year. But this type of nonstationarity is irrelevant for the issue at hand. What counts is whether the model's central parameters—in particular, the probability that the drug is good or bad—are nonstationary. And this cannot be established merely by pointing to substantive changes in drugs over time. (For example, because the pharmaceutical firms typically operate on the frontier of R&D knowledge, the underlying probability that new drugs are both safe and effective may stay roughly constant for long periods of time. That is, the substantive content of the R&D frontier changes, but the corresponding risks of working on that frontier—represented as success probabilities—need not.) Lastly, what is considered "stationary" partly depends on what is being compared to what. The FDA makes many decisions on new drug applications every year. If the process of drug approval is relatively quick, compared to how fast parameters such as $p_g$ change, then over an empirically meaningful time frame the assumption of stationarity is a reasonable first approximation.

## Optimizing Is Difficult

It is virtually a folk conjecture among formally inclined political scientists that in a stationary environment, many kinds of sensible adaptation converge to optimal behavior. Theorem 1 indicates that this view is overconfident: adaptive schemes that satisfy (A1)—a sensible property—fall short of optimality even in the long run. We suspect that this overconfidence arises from a failure to appreciate how hard it is to optimize when the effectiveness of different options is unknown. This uncertainty creates a fundamental tension between exploration and (wise) persistence; resolving this tension optimally is, in general, difficult.[25] To give a sense of how hard it is, reconsider the famous Bush-Mosteller rule. Because the speed of learning in this rule is a fixed parameter (i.e., it is independent of time), the canonical Bush-Mosteller rule satisfies (A1). Hence, a person who adapts via this scheme will not settle on his or her optimal action if it is imperfect. As with any (A1)-type rule, the culprit is relentless learning, driven by an implicit refusal to tolerate any error. A natural question arises: is it possible to dampen this restlessness of the Bush-Mosteller approach by making the speed of learning time dependent? (Note that corollary 1 tells us that we must do so since optimality requires violating [A1].) The answer is yes— *but it is easy to overdo this dampening* and make the scheme too sluggish. Specifically, one can show that performance is improved if $\alpha_t$, the (now time-dependent) speed of learning, $\simeq \frac{1}{t}$ (Robbins and Monro 1951). However, $\alpha \simeq \frac{1}{t^2}$ is overkill. It is not intuitively obvious why $\alpha_t \simeq \frac{1}{t}$ is just right, whereas $\alpha_t \simeq \frac{1}{t^2}$ is too much dampening.[26] Thus, we see that getting the trade-off between exploration and search just right is no mean feat. (This is why Robbins and Monro's paper was considered a major breakthrough.) Indeed, it is a delicate matter. There is no compelling reason to believe that many garden-variety heuristics have this unusual property.

### Bad Normative News May Be Good Methodological News

That an adaptively rational agency of this type cannot learn how to make decisions optimally, even when it has unlimited time and confronts a stationary task environment, sounds like bad news. So it is, in terms of our normative assessments of what adaptive behavior can achieve for us. But in another respect—empirically discriminating between rival theories—it is good news.

After all, we can regard the preceding model as a purely empirical theory of organizational choice, setting aside its normative implications. Let us contrast it with an increasingly popular family of theories, rational choice models of bureaucracy, where these are also considered to be descriptive theories. (See Weingast 1996 for an overview of such models.) Suppose that the choice problem is nontrivial, from the bureau's perspective.[27]

Thus, the premise of an optimizing bureau is not equivalent to assuming that the agency maximizes some measure of social welfare. Instead, the weaker assumption—the bureau maximizes *some* utility function—applies here. Social welfare could be a component of that utility function, but it need not be.[28]

With this caveat in mind, the claim that a completely rational agency will be optimally reliable (i.e., on its own terms) should cause no confusion. The latter follows immediately from the former.

And since a fully rational agency is optimally reliable, theorem 1 tells us that we have a real horse race between this type of rational choice theory and our adaptive theory. Viewed as descriptive theories, the two clearly compete with each other: they offer different predictions about the agency's behavior. In a stationary environment, a rational agency that knows (or learns) the relevant probabilities ultimately fixes on a unique decisional standard; an adaptive agency that adjusts via any (A1)-type rule does not. We suspect the evidence tilts toward our theory.

### Other Extensions of the Reliability Model

The substantive problem above centers on the trade-off between type I and type II errors. But the model can be extended to cover other kinds of reliability problems. Two other situations are worth noting here. Each case is an instance of classical reliability theory in that a system exhibits only a single type of reliability—the system is either working or it isn't—unlike the two types of reliability (or errors) of FDA-like systems.

(1) Suppose a manager is responsible for two ongoing projects. These projects are technologically independent, but they share a common resource constraint: the manager's attention. Hence, the decision variable is the fraction of time the manager devotes to each project. The more time spent on a project, the more reliable it is. A putting-out-the-fire response would be to increase the time he or she spends on the project that has most recently exhibited reliability problems. Then it is easy to show that there is again a difference between optimal and adaptive reliability: if the manager's problem of allocating time is nontrivial, then he or she never converges to the optimal allocation, paralleling theorem 1's part (ii).

(2) Suppose a manager is responsible for only one project but is also held accountable for its cost. Spending money tends to improve reliability, and the manager puts out fires by increasing spending on reliability after something has gone wrong. But because running a tight ship is also an organizational goal, the manager has a tendency to cut costs when all is quiet (Perrow 1984).[29] Then typically there will exist an optimal level of spending on reliability, dictated by variables such as decreasing returns to the value of the $n$th redundant unit; but once again, the manager's ceaseless adaptation makes stabilizing at this optimal level impossible. Instead, the organization will oscillate between too little and too much redundancy, sometimes pausing—but never staying—at the optimal amount.

These two extensions show that the fundamental element in the current analysis is not the trade-off between type I and type II errors but the difference between certain kinds of adaptive reliability—reflected by rules involving never-ending adjustment—and optimal reliability. Indeed, the search for optimal *reliability* is not central at all; rather, the key is the pathology of excessively high aspirations, the unwillingness to settle for imperfect-yet-optimal decision rules.

Hence, this investigation into organizational reliability taps into very broad issues: it is related to the long-standing controversy in the social sciences, triggered by Simon's Nobel prize–winning work (1955, 1956) on whether we should model individual actors as boundedly or fully rational.[30] Some have argued that this debate is irrelevant because both types of actors

ultimately make the same decisions in equilibrium. Others have argued
that imperfectly rational agents do not make the decisions that completely
rational ones do, even in the very long run. This chapter clarifies this debate
by establishing that certain plausible kinds of adaptive rationality do *not*
converge—even in the long run—to optimal behavior (see also chapter 3).
Hence, we cannot be content with vague claims, supported by loose arguments, about the relation between limited and complete rationality: exactly
*how* rationality is bounded matters. The debate will be productive only if
we examine well-specified forms of adaptation and then *prove* that indeed
they do, or do not, lead to optimality. (See Conlisk 1996 for a cogent
argument along these lines.)

This chapter focuses on the substantive problem of organizational reliability, which—as recent events attest—is important in its own right. On this
subject, we think that the last word belongs to Landau and Chisholm. In
"The Arrogance of Optimism," they cautioned us that when organizations
are created, "the potential for error is inescapable. Fault-proof organizations are beyond our design capability" (1995, p. 69). We agree completely:
it is the height of arrogance to presume that one has designed an error-free
system. Every real system, be it hardware or an organization, sometimes
fails. Even the *best possible* system—the optimal one—displays errors. Failure to appreciate this fundamental fact leads to an interesting phenomenon:
*excessive adaptation.*

The beginning of wisdom, in the study of organizational reliability as
elsewhere, requires a grasp of what is possible.

## CHAPTER 6

# Garbage Can Theory

JONATHAN BENDOR, TERRY MOE,
AND KEN SHOTTS

## INTRODUCTION

Chapters 1 and 2 suggest that the research program of bounded rationality (BR) has two main branches: Simon and Lindblom's approach, which focuses on how adaptive we can be despite our cognitive constraints; and Kahneman and Tversky's, which emphasizes how easily we can err even when confronted by simple problems. There are, however, other approaches to decision making that some scholars believe belong to the BR program. Prominent among these is garbage can theory (Cohen, March, and Olsen 1972), hereafter GCT. Although GCT's links to BR are not completely straightforward, neither are they completely opaque. In particular, the theory's emphasis on organizational technologies that are unclear to the very decision makers who use those technologies makes it a candidate for membership in a research program defined by cognitive constraints.[1]

Thus, this chapter takes seriously the claim that GCT is part of the BR program. It also takes the GCT itself seriously by reproducing a sustained critique (Bendor, Moe, and Shotts 2001) of work in the garbage can tradition. We turn now to that critique.[2]

## RECYCLING THE GARBAGE CAN

GCT has led a charmed life—to its great disadvantage. Although proponents have effectively advanced their ideas, potential critics have done little to bring about a productive intellectual exchange. This one-sidedness impedes scientific progress. And it surely has hurt the GCT, which, despite its prominence and despite the optimism and enthusiasm that mark its history, has never been able to overcome its inherent weaknesses and make the kind of contribution that once seemed its destiny.

This is unfortunate. Cohen, March, and Olsen's first paper is brimming with provocative insights that offer a promising basis for theory, but thus far their potential has largely gone untapped. Indeed, the theory that has grown up over the years is so complex and confusing, and some of its components are so seriously flawed, that there is little reason for thinking that it can look ahead to a more fertile future. For fundamental reasons, the theory lacks the rigor, discipline, and analytic power needed for genuine progress. We want to explain why this is so and what might be done to get it back on track. Why expend so much energy examining an old article and its progeny? Such an exercise would be eminently worthwhile if only because some of the ideas are well worth salvaging. But there is even more to be gained. The 1972 article has had considerable influence on political science and on institutional theory more generally.[3] Some of this influence is directly due to the original article, whose ideas have shaped thinking about organizations ever since and which presents one of the most famous computer simulations in all of social science. Perhaps most of its influence, however, is now exercised indirectly, through more recent works that have moved into new terrain and gained much attention in their own right but are grounded in the original theory. A few of the latter have been carried out in the work of other authors, notably Kingdon's *Agendas, Alternatives, and Public Policies* (1984), one of the most highly regarded treatments of policy making in the past two decades. But virtually all the basic theoretical work in this tradition has been done by the original authors, particularly March and Olsen.

Consider the lineage. *Ambiguity and Choice in Organizations* (March and Olsen 1976a) expanded on the 1972 theory, received considerable attention, and solidified the garbage can's new status as a major school of organization theory. Subsequently, in 1984, the *American Political Science Review* published March and Olsen's hugely influential overview article, "The New Institutionalism," which officially ushered the new institutionalism into political science and gave a prominent role to the garbage can and its progeny in institutional theory. Henceforth, political scientists would routinely regard the March-Olsen line of theory as a main variant of the new institutionalism (Peters 1996). This article, in turn, developed into March and Olsen's longer statement on the new institutionalism, *Rediscovering Institutions* (1989), which, as we will show, was strongly influenced by the original garbage can theory and its extensions. This book met with much acclaim and has been judged a "contemporary classic" by Goodin and Klingemann (1996) in their survey of the discipline.

This is an impressive hit parade; clearly, the GCT has had an important impact on the field. Yet, remarkably, there have been no systematic attempts to assess the theory or even to clarify its logic and content so that everyone can agree on what it actually says. This is not true for the two major approaches to institutional analysis: theories of rational choice (e.g., Weingast 1996) and of bounded rationality (Cyert and March 1963; Simon 1947; March and Simon 1958). Scholars are quite clear about what these theories say, and both traditions have been the subject of debate and criticism (e.g., Green and Shapiro 1994). The garbage can theory has escaped all this. Even if it were a shining example of social science theory, this would be unhealthy, for no theory and no field of study can do without criticism, clarity, and debate. The situation is all the more worrisome given the intellectual influence the garbage can has had over the years. If we are right about the theory deficiencies, this influence cannot help but propagate problems that the discipline should avoid in its pursuit of truly productive theories.

We offer a critical assessment of the garbage can theory that is as comprehensive as possible, befitting a research program that has gone unchallenged for its entire lifetime. We also seek to clarify what the theory actually says, which is absolutely necessary given the pervasive confusion that surrounds it. Finally, we try to map out the theory's intellectual lineage to show how the original work—and its problems—are reflected in the new institutionalism. This is a lot to do in one article; there is not enough space to do more, although we would like to. Our immediate goal is to stimulate an intellectual exchange that will promote a better understanding of the garbage can approach and its role in institutional theory. In the longer term, we hope to be part of a collective effort to reconstruct the theory around some of its more promising (and still exciting) ideas and thus help revitalize the research program and set it on a more productive path.

We cannot delve into the details of this reconstruction here. That remains for future work. But we do think—for reasons we will cover below—that the garbage can's key ideas can best be developed by anchoring them firmly in theories of rational choice or bounded rationality, most likely the latter, whose analytic apparatus for exploring decision making under cognitive constraints is particularly appropriate. Understood in this way, a revised garbage can theory would become a parametric variant of the bounded rationality program. This would clarify the garbage can's theoretical logic, give it a clear microfoundation, and provide it with real deductive power for pursuing its insights—all of which it now lacks.

## THE VERBAL THEORY AND THE COMPUTER MODEL

What is the garbage can theory? This is a simple question with no simple answer. The original article sets out two distinct formulations. The first is a set of ideas about how decisions get made in "organized anarchies." These ideas, because they are expressed in ordinary English and their logical relations are not rigorously spelled out, constitute a verbal (or informal) theory. This is Cohen, March, and Olsen's (1972) general statement of the garbage can theory as originally conceived. The second formulation is the computer model, which is intended to capture in simplified, mathematical form the verbal theory's key features and to provide a rigorous logical structure that can generate predictions. We will examine both of these.

The two formulations should be kept separate. The verbal theory is fundamental. The computer model is derivative: it is only one of many possible ways to formalize key features of the theory and draw out its implications. But we must remember that no formalization is unique. We must also remember that the sophistication and rigor of the simulation do not in themselves guarantee that it faithfully reflects the theory; depending on the details of its design, so much may be lost in the translation that the model could have little value or even prove misleading.

The place to start if we want to understand the GCT, then, is the verbal theory. The original article, however, sheds surprisingly little light on its content. Only two pages discuss it, and little is done to develop, connect, or clarify its central ideas. Instead, almost all the article is devoted to the computer model, whose properties are discussed in detail and whose implications are discussed at length and then applied to a specific empirical domain (universities).

Most of what we know about the informal theory comes not from the original article but from later work, particularly March and Olsen's *Ambiguity and Choice in Organizaions* (1976a), March and Weissinger-Baylon's *Ambiguity and Command* (1986), and March and Olsen's *Rediscovering Institutions* (1989). (See also March's *A Primer on Decision Making* [1994].) This later literature does more than fill in the basic contours of the original theory, however. It also dramatically extends the scope and complexity of garbage can–like thinking and ultimately develops a new theory, the March-Olsen variant of the new institutionalism, which makes frequent, although imprecise, use of GCT ideas.

Unfortunately, the authors and their colleagues rarely distinguish between the informal theory and the computer model in discussing GCT and what it tells us about organizations. This holds for most of the more recent literature on the GCT and is even true of the original article. There the authors begin with a clear distinction between the two, but by the concluding section they have tossed both into the same idea pool, characterizing garbage can processes by properties that come sometimes from the verbal theory and sometimes from the computer model—but without any recognition that they are doing so. GCT thus emerges as an undifferentiated blend of the two.

We will sort through all this by proceeding as follows. First, we clarify and evaluate the ideas that appear most central to the verbal theory, emphasizing those in the original article but including some introduced in more recent work when appropriate. Second, we turn to the computer model, clarify its structure and operation, and assess its contribution to the larger enterprise. Third, we examine GCT's post-1972 theoretical development, including its influence on the March-Olsen variant of the new institutionalism, and assess these newer lines of research. Finally, we briefly illustrate how the GCT might be revitalized by using some classical ideas of bounded rationality to analyze one of the garbage can's most provocative insights.

## THE EARLY VERBAL THEORY

The verbal theory is about organized anarchies, a class of organizations or decision situations marked by problematic preferences, unclear technology, and fluid participation. As the definition suggests, organized anarchies face certain kinds of ambiguity. This is what makes them interesting. GCT is an effort to explain how organizations make choices and solve problems under conditions of ambiguity so troubling that they appear to render decision making extremely difficult or impossible.

The various branches of rational choice theory have long had sophisticated tools for modeling individual choice under uncertainty. But GCT is founded on a radically different approach to choice. Its premise is that choice in an organized anarchy cannot be understood via the intentions of organizational participants, and imposing a rational explanation on organizational behavior can only distort what is really going on. Choices often just happen, with no clear connection to what participants want. They arise from dynamic organizational processes that are complex, highly contextual, and unpredictable, driven more by accident and timing than by individual intention. This gives much of organizational life an almost chaotic appearance.

At the heart of GCT is the notion that organizational outcomes arise from independent "streams" of problems, solutions, participants, and choice opportunities, whose random intersection generates decisions. In this scheme, choice opportunities are the garbage cans. As problems, solutions, and participants move independently about the organization, various combinations find themselves dumped into these cans, and whatever decisions come out depend on the mixtures the intersecting streams happen to generate. The organization's basic structure—notably, its rules—channels and constrains this multifarious action, thereby shaping the organization's patterns of choice and problem solving. But the driving force of the garbage can explanation is process.

The process-driven world of the garbage can is more than dynamic. It is also strange and even pathological by conventional standards. Alice has gone through the looking glass, and nothing is as it seems. Choices happen for no apparent reason. Outcomes are divorced from intention. Solutions are disconnected from problems. People wander aimlessly in and out of decision arenas. In this welter of loosely coupled activity, some decisions do get made, and some problems do get solved. There is a chaotic brand of "intelligence" at work. But it is largely due to the time-dependent confluence of events, not to the rational effects of plans or goals.

The organizational literature often calls the GCT a metaphor. Even the authors have used this term (e.g., March and Olsen 1986). In a sense this rings true: the notion of different inputs being mixed together in a garbage can, with output depending on what is mixed with what, is clearly metaphorical. But we must be careful with this kind of language, because calling a set of ideas a metaphor may inappropriately shield it from criticism. A metaphor is, after all, a literary device that need not meet the same standards as social science theories. Yet the GCT's basic ideas are obviously intended to be a theory and should be treated as such.

Any effort to clarify and appraise the theory could highlight a large number of issues. We think the following are among the most important.

### Individual Choice and the Level of Explanation

GCT is a theory of organizational choice. In developing it, the authors naturally move back and forth between individuals and organizations. They discuss how individuals make choices and how these choices depart from standard notions of rationality. They also talk about how, as an organization's dynamics work themselves out, the choices of individuals are aggregated into organizational choices. The impression is that GCT explains organizational choice via a logically coherent, if unorthodox, theory of individual choice.

But this is not really so. Although the authors make clear that they are abandoning conventional rational choice approaches, they do not develop their own model of individual choice that can function as a component of GCT. Indeed, the original article offers no theoretical treatment of individuals at all; and in the computer model individuals are little more than automatons, barely touched by motivations or beliefs. The article fashions a theory of organizational choice based on assumptions about process and structure—but without any underlying theory of individual choice. The explanation is at the macrolevel.

In more recent work, which moves toward a larger theory, the authors do talk theoretically about individuals and offer arguments about how decisions are made in the face of ambiguity. In these discussions, however, everything inherent to individual choice is left endogenous to the choice process. Among other things, it is not assumed that preferences exist prior to choice or that they determine actors' evaluation of alternatives: rather, people discover their preferences only after they have decided, which begs the question of how they made the decisions in the first place.

Empirically, it might seem attractive to relax every assumption of classical rational choice theory and argue that all these aspects depend on experience. Analytically, however, with everything treated as endogenous and nothing taken as given, there is no logical basis for deriving behavioral expectations or for moving from individual to organizational properties. The authors try to fill in the gaps by referring to other possible foundations of individual behavior—symbols, rules, duty, obligation, myths, and the like—but these complications, which may or may not be empirically warranted, do not add up to a coherent model of individual choice that can generate testable implications. If anything, they hinder a serious effort to construct one.

However one might evaluate these more recent efforts to build a theory of individual choice, the early GCT is essentially a macrotheory of organizations. It talks a lot about choice and individuals, but the theory really focuses on process and structure—organization-level phenomena—and does not arise from individualistic foundations.

## Independent Streams

The hallmark of GCT is its attempt to understand organizations by reference to streams of problems, solutions, participants, and choice opportunities, which are independent of one another and exogenous to the system. On the surface, this formulation seems to be a simple, straightforward way of characterizing the internal dynamics of organization. It is memorably provocative, widely cited—and crumbles upon closer examination.

First, in the real world problems are typically identified and put on organizational agendas by organizational participants. Similarly, solutions find their way into organizations because they are pushed by certain participants in light of the problems they face. There is nothing mysterious about this. It is obvious. *People are the carriers of problems and solutions.* Indeed, when Cohen, March, and Olsen discuss garbage can processes, they often say exactly that. Their original article (1972) even included this point in its list of "basic ideas": "To understand processes within organizations, one can view a choice opportunity as a garbage can into which various kinds of problems and solutions are dumped *by participants* as they are generated" (p. 2, emphasis added). This sort of reasoning is repeated elsewhere. But if problems and solutions are dumped by participants, then how can streams of problems and solutions be independent of the stream of participants? Clearly, which problems and which solutions get dumped into a particular garbage can depend on which participants happen to go there. The streams are not independent.[4]

There is a second important source of confusion that surrounds the notion of independent streams. This one arises from an odd use of language. Take, for instance, the concept of "solution." *Webster's Collegiate Dictionary* (9th ed.) defines this term as "an action or process of solving a problem," which is the way it is commonly used and understood. By definition, then, solutions do not exist on their own, independent of specific problems. Nor does it make sense to call an object (like a computer) or an idea (like decentralization) a solution without indicating what problem it is meant to address. Objects are objects. Ideas are ideas. They qualify as solutions, and we are justified in applying that concept to them, only when we show that they somehow address particular problems. It follows that streams of solutions and problems cannot possibly be independent of one another—unless commonly accepted definitions are abandoned.

The 1972 article defines a solution as "somebody's product" (p. 3), using a computer as an example. By this definition, there is no logical connection between solutions and problems, and the two can be decoupled into independent streams. GCT can then talk provocatively about solutions chasing problems and problems chasing solutions, a standard ingredient of garbage can chaos. But all this turns on a definitional distortion. GCT's concept of solution is not the one everyone else uses; it is the same word invested with an entirely different meaning. Hence, the theory is not about solutions as we understand them. It is about something else. Had this something else been given a label of its own, this source of confusion would have been eliminated. But GCT would have appeared much narrower and less interesting, and its claims, of course, would have been very different.

This linchpin of GCT's framework, then, does not stand up to scrutiny. The notion of independent streams may appear at first blush to be an inspired simplification, but one is hard put to see how it can be justified.

## Organizational Structure

The original article emphasizes organizational dynamics, but it also recognizes a significant role for structure, which influences how people, problems, and solutions get matched up in choice situations and what outcomes are generated as a result. What it says about structure, however, is very limited—and what it leaves out is crucial.[5] Two problems stand out.

The first arises because the early GCT takes structure as exogenous, thus ignoring where structure comes from or why it takes the forms it does. The theory's focus is on the effects of structure, not on its origin, design, or change. All else equal, we would not object to this. Something must be taken as exogenous; not everything can be explained. But in this case there is a heavy price to pay.

Precisely because structure does have a major effect on organizational processes and outcomes—and therefore on its effectiveness and its distribution of costs and benefits—decision makers are likely to regard it as a key means of shaping organizational behavior.[6] This is unavoidable. To say that structure has important effects is to imply that it can be used to achieve certain ends. And participants, especially those in authority, have incentives to use structure in this way. When leaders act on these incentives, then, they need not accept the garbage can's chaotic dynamics as uncontrollable facts of life. If GCT is correct in claiming that garbage can processes work very differently in different structures, then the theory's own logic suggests that leaders could choose structures with this in mind.

It is difficult to deny that the instrumental use of structure by leaders is pervasive and fundamental to an understanding of organizations. Yet GCT essentially ignores it. The result is that GCT is a theory of short-run dynamics that fails to account for how structure would adjust over time to garbage can processes flowing within—and it thus fails to explain how organized anarchies would ultimately behave. The thrust of GCT is to highlight the chaos, perversity, and ineffectiveness of organized anarchy; what the theory omits are the structural adjustments that could create more orderly and productive organizations. Absent structural choice, the deck is stacked.

The second problem with the article's treatment of structure concerns the kinds of structure considered relevant to an understanding of organization. Essential components of real organizations are simply omitted and thus can never figure into the garbage can's short-term dynamics. The result is a perspective that misses and mischaracterizes much of organizational behavior.

The most significant omissions bear on issues of authority, delegation, and control. Whether or not organizations have neat hierarchies, virtually all—including universities, the quintessential organized anarchies–have authority structures of some sort. In these systems, superiors can tell others what to do, hire subordinates to do their bidding, and delegate authority to them. They thereby create control opportunities: they can get much more done by extending their reach into far-flung decision arenas, even if they cannot personally participate. But they also create control problems: subordinates will not necessarily do what is in the superior's best interests. Other structures, such as incentive schemes and monitoring systems, are used to promote as much congruence as possible. The structure of an organization largely reflects efforts to control and coordinate the actions of interdependent individuals.

These issues are at the heart of organization and have long been at the core of organization theory. GCT, however, is built around a simplified framework of independent streams that largely ignores them.[7] Although the authors recognize that some participants may be more important than others, in having more access to important decision arenas, they make no systematic attempt to recognize that participants are connected—notably, that many are agents of others—and that these connections are not accidental: they are designed to extend the reach of superiors. Instead, the world of the garbage can is made up of anomic individuals (the notion of a "stream" of participants says it all), and organizational decisions turn on who happens to wander into particular decision arenas.

Structure can induce order and productivity, but GCT does not allow participants to recognize as much and simply removes structure from their realm of choice. Were these issues of structure seriously integrated into the theory, it is hard to see how the garbage can's basic claims could be maintained.

## Organized Anarchy

A fundamental GCT claim is that in a certain domain, organized anarchy, garbage can processes prevail. To test this hypothesis one must identify the domain. What, then, are organized anarchies (OAs)? The 1972 paper defines them as "organizations characterized by problematic preferences, unclear technology, and fluid participation" (p. 1). But this definition, though widely cited, suffers from serious ambiguities that subsequent work has done little to clarify.

The first problem arises from the 1972 article's ambiguous statements about whether OAs must have all three properties or just some of them. Much of the discussion suggests that each property is necessary.[8] Yet, in the conclusion, OAs are described as situations that depart from classical decision models "in *some or all* of three important ways: preferences are problematic, technology is unclear, *or* participation is fluid" (p. 16, emphasis added); that is, each property suffices to identify an organized anarchy.

Confusing the logic of "and" versus "or" creates two problems. First, the size of the concept's domain is unclear: it is much larger if only one property is needed to identify an OA than if all three are required. Second, it opens up a theoretical can of worms for the research program. For instance, under the "some or all" stipulation, an agency with clear goals, rigid participation, and unclear technology and one with clear goals, clear technology, and fluid participation would both be OAs. But why would we expect either to behave like garbage cans? In both cases preferences are unproblematic, so solutions could not chase problems; options that fail the (clearly) relevant criteria will simply not be considered solutions. Furthermore, what are the theoretical reasons for assigning both agencies to the same set? Would not an organization with unclear technology and rigid participation behave differently from one with clear technology and fluid participation?

Finally, the meaning of each property in isolation is open to question. Consider universities, which are held up as paradigmatic organized anarchies. In what sense is the technology of teaching unclear? True, no one knows how to teach graduate students to become creative researchers. But we do know how to teach statistics, chemistry, and many other subjects: well-known sequences of readings, lectures, problem sets, and discussion sections produce a stable percentage of students who grasp the material, as verified by exams. Thus, vital aspects of teaching are understood rather well. So is this an unclear technology or not? The garbage can literature gives us no guidance. Much the same could he said for problematic preferences and fluid participation.

## GCT and the March-Simon Tradition

It is a common perception (e.g., Goodin 1999, pp. 71–72) that GCT, though now important enough to be considered a research program of its own, evolved from the earlier behavioral theories of March and Simon and thus can be considered an offspring of the decision-making school of organization theory. On the surface, this makes sense. After all, James March is a founder of both the Carnegie school and GCT, so one might expect a strong continuity of ideas. Moreover, Cohen, March, and Olsen go to some lengths to describe their work in exactly this way, emphasizing that they "build on current behavioral theories of organizations" (1972, p. 1), that their ideas about organizations "have a broader parentage" (p. 2) in that literature, and that their intentions are "conservative," aiming "to continue a course of theoretical development" with roots in the decision-making tradition (March and Olsen 1976c, p. 22).

With effort, one might make a case for this view. On the whole, however, it mischaracterizes the garbage can. The March-Simon tradition is grounded in a theory of individual choice. Although it relaxes classical rational choice assumptions to account for human cognition, it borrows heavily from economic methodology (e.g., Cyert and March 1963), and its explanations contain a strong individualistic component (e.g., March and Simon 1958, chapter 6). It sees behavior as intendedly rational, it sees organizations as products of boundedly rational choice, and it seeks to model all this as clearly and simply as possible.

At the heart of the March-Simon tradition is an analysis of how cognitively constrained individuals solve problems and what this entails for organization. The theory forges a connection between bounded rationality and organizational structure, demonstrating that, precisely because individuals are limited in knowledge and computational abilities, they must rely on organizational routines and other forms of programmed behavior to solve their problems (Simon 1947, pp. 79–80, 88–89). Above all, the theory provides an explanation of structure, arguing that it arises out of the cognitively limited yet intendedly rational behavior of very human problem solvers.

The garbage can's approach is very different. Three contrasts stand out.

(1)  Most fundamentally, GCT rejects the key tenet of the entire Carnegie tradition: that individuals are intendedly rational. Indeed, it does not even portray individuals as caring about solving problems.

(2)  GCT does not relax rational choice assumptions selectively, as March and Simon do, in order to retain some of the power and rigor of the methodology. It rejects all the relevant choice-theoretic assumptions and abandons the entire program of methodological individualism. In contrast, what the March-Simon tradition says about organizations is firmly rooted in a coherent model of individual choice. That is its analytic foundation.

(3)  The basic aspects of organization that March and Simon seek to explain—structure (especially) and process—are treated by GCT as exogenous. No explanation is given. The aim of GCT is to explain outcomes, not organization. The aim of the March-Simon tradition is to explain organization.

We must not confuse GCT with its authors, who have surely been influenced by the March-Simon tradition. GCT itself, however, is only tenuously connected to that tradition; indeed, it rejects the decision school's core components. So it is misleading to think of the current formulation of GCT as an outgrowth of bounded rationality conceptions of organization. A close look at the theory suggests that it has much more in common with the institutional school of organization theory (e.g., Meyer and Rowan 1977), which also rejects the idea that organizations can be understood in terms of the rational behavior of their members. Institutional theory emphasizes other influences—symbols, myths, legitimacy—that are reflected, nonrationally or even irrationally, in the foundations of organization. As the GCT authors develop their larger theory, it is evident that this is the branch of organization theory where the garbage can is most at home.

## THE SIMULATION MODEL

The computer simulation formalizes the verbal theory and is widely regarded as the research program's scientific core. We agree that formal modeling is desirable, but the specific model the GCT authors developed misrepresents the theory and exacerbates the GCT's problems. To show this, we need to describe the model's basic assumptions and how they generate outcomes.[9]

The computer model simulates the behavior of an organized anarchy over twenty periods, in which the various organizational streams interact to produce outcomes. The choice stream consists of ten choice arenas (the garbage cans), the participant stream of ten decision makers, and the problem stream of twenty problems.

This much follows the informal theory. But notice, right at the outset, what is missing. Whereas the verbal theory posits a central role for streams of choices, participants, problems, *and* solutions, *the computer model has no stream of solutions.* No explanation is ever offered of why this stream is missing. Given the theory, the temporal pairing of problems with solutions ought to be fundamental. And the notion of "solutions chasing problems and problems chasing solutions"—perhaps the most famous property of the verbal theory—cannot be part of the model if the simulation does not include solutions in the first place.

Let us put this point aside for now, however, and return to describing the computer model as it was constructed. The temporal sequencing is as follows. During each of the first ten periods, one new choice arena and two new problems enter the simulation, where they may be acted upon by the organization's decision makers, all ten of whom are assumed to be present from the start. By the tenth period, all choices and problems have been introduced. There are no new inputs during the last ten periods.

During each period, each problem and each decision maker attaches itself to (at most) one choice arena. Each decides where to go, subject to two constraints. First, no problem or agent can go to a choice that has not yet been introduced or has already been made. Second, the options are limited by the organization's access structure (for problems) and decision structure (for participants).

The access structure determines which problems have access to which choice arenas. In an unsegmented access structure, the problems can go anywhere: They can attach themselves to any active choice arena. In the hierarchical structure, problems are ranked by importance: The two most important problems can go to any choice arena, the next two can go to nine of the ten choice arenas (the off-limits arena being too important for them), and so on, down to the two least important problems, which can go only to one of the ten arenas. In the specialized access structure, each problem can go only to one particular choice arena: two of the problems can go only to arena 1, two can go only to arena 2, and so on.

Similarly, the decision structure determines where decision makers can go. In the unsegmented decision structure, any agent can work on any choice. In the hierarchical structure, important decision makers have access to more choice arenas than do less important ones. And in the specialized structure, each agent can work in only one arena.

Within these constraints, each problem and decision maker follows a simple rule in selecting a choice arena: go to the choice that is closest to being made. This rule is a crucial assumption that drives many of the results. We will discuss how it works in more detail below, after describing how choices are made.

### How Choices Are Made

In the model, the key concept for understanding how choices are made is "energy"—a concept that the informal theory does not use. The model, unlike the theory, sees choices as the product of an energy-balancing mechanism. It works as follows.

Each problem requires a certain amount of energy to be solved. This requirement is the same for all problems in an organization but can vary across organizations depending on the load they face. (The higher the load, the more energy is required per problem.) The energy required to make a choice is the sum of the energy requirements of the problems attached to that arena.

Decision makers supply the energy that gets things done. In each period a participant supplies a fixed amount of energy to the choice on which he or she is currently working. However, not all this energy is assumed to be usable in solving problems. The usable energy is derived by multiplying the available energy by a fraction, set in the model at 0.6. The authors interpret this fraction as representing "solutions" in the simulation. Presumably, larger fractions mean more usable energy and thus more solution power. To reiterate, however, there is no stream of solutions, as the verbal theory requires. Indeed, apart from this fixed fraction, solutions play no role in the model.

In any period, the energy supplied to a choice arena is the sum of the amounts brought by each of the participants to that arena. Curiously, this energy supply is assumed to cumulate over time: the total energy in a choice arena is the sum of all the energy anyone has ever brought to that arena since that choice was first introduced. This creates the possibility of "ghost" energy (our term): a choice arena may have usable energy in a period even though no one is in the arena at that time.

If the total usable energy equals or exceeds the energy required by the problems currently attached to the arena, then the choice is made and the problems are solved. Because of ghost energy, however, problems may be solved even though no one is present to work on them. More perplexing still, a choice can be "made" *even when there are no problems in the choice arena*. Indeed, this is likely, because when there are no problems, the energy requirement is zero, and the presence of any agents or (even without participants) any ghost energy automatically satisfies the energy requirement. By the simulation's logic, then, a choice can be "made" even though no problems are solved, no one is present, and, quite literally, nothing happens.

This is not, of course, what is normally meant when we say a choice has been "made." To claim, as the model does, that a choice is made when nothing is happening is an odd use of language and ultimately a source of serious confusion. It is better to say, under these circumstances, that the choice arena closes down, or the meeting adjourns without any decisions being taken. This accurately describes what actually happens, and it does not imply that choices are being made when they are not. We will return to this peculiar form of decision making below, for it plays an important part in GCT analysis.

Finally, we want to be clear about how energy relates to the movement of people and problems. As noted, both follow a simple decision rule: go to the choice that is closest to being made. Reinterpreted in energy terms, this rule says, go to the choice arena with the smallest energy deficit—that is, the smallest gap between the energy required to solve problems and the energy supplied by agents (including ghost energy). In each period, problems and participants look over all choice arenas, calculate each arena's energy deficit, and move in the next period to the open arena with the smallest deficit (subject to the constraints of the access and decision structures).

## The Verbal and Formal Theories: Different Worlds

The elements we have discussed are the model's essential components, which largely determine how it simulates the behavior of organized anarchies. We have already pointed out some serious design problems: the absence of a solution stream, for instance, and the notion that choices can be "made" even when no people or problems are present. Before we go on, we want to add four general observations.

First, the informal version of the GCT is clearly an information-processing formulation. It refers to intelligence in the face of ambiguity, technologies that are poorly understood, and the allocation of attention. All these ideas are fundamentally cognitive or informational. The simulation is markedly different. It is built around the concept of energy and does not explicitly represent information or the beliefs of decision makers. Instead, participants are simply carriers of energy. Further, problems are defined in terms of energy requirements, solutions are coefficients that convert available energy to usable energy, people and problems move around in response to energy deficits, and choices are made when energy deficits are overcome. Nothing here resembles the verbal theory.

Second, the model's assumptions about participants are especially odd. Decision makers are basically bundles of energy—virtually automatons. They have no explicit objectives and are essentially unmotivated. They do follow a decision rule: they go to choices that are closest to being made. But the authors do not say that this reflects an underlying goal, nor do they explain why decision makers might want to do this. We do know that participants do not care in the slightest about solving problems. They are perfectly happy to go to an arena that contains no problems and participate in the "decision." Indeed, because such a can has a light energy load, their decision rule implies that they actually prefer it to one with problems. It is hardly surprising, then, that the simulation ultimately shows that organizations are bad at solving problems. No one is interested in solving them.

Third, the verbal theory is about a stochastic world filled with ambiguity, chaos, and randomly intersecting streams of activity. Its depiction of organizational dynamics is built around randomness and the uncertain confluence of events. But the simulation is fundamentally deterministic. People and problems move at the same time, using the same deterministic rule. And the technology assures that choices will always be made in the same way when certain energy requirements are met. The world created by the computer simulation is dynamic—people and problems move around over time—but it is clearly not the stochastic world the informal theory describes because it omits the essence of what that world is supposed to be about.

Fourth, if we accept the interpretation that the three properties of problematic preferences, unclear technology, and fluid participation are individually necessary and jointly sufficient for something to be an organized anarchy, then this central concept of the informal theory is poorly represented by the simulation. Compare these defining properties of an OA to the simulation's nine structural variants, shown in figure 8.[10]

When the access structure is hierarchical (versions corresponding to cells 4, 5, and 6), problems are ranked by importance. Thus, the organization's preference ordering is completely coherent; its preferences are not problematic at all.[11] Regarding technology, versions in cells 7, 8, and 9 have

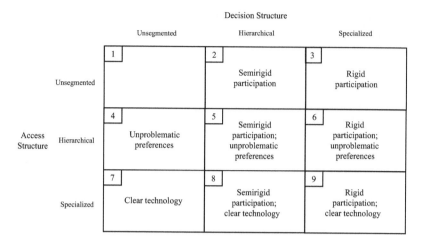

Note: The decision structure maps decision makers onto choice arenas. The access structure maps problems onto choice arenas. Each cell represents a variant of the simulation.

Figure 8. The simulation's nine structural variants.

specialized access structures: problems are coded by type and go to appropriately coded choice arenas. This is a clear technology: the organization knows enough about problems to categorize them correctly and route them accordingly. Finally, participation is completely rigid in cells 3, 6, and 9, for these have specialized decision structures that restrict each agent to a single choice arena; it is semirigid in cells 2, 5, and 8, where lower-level decision makers are sharply constrained. Thus, to remain true to the verbal theory's central ideas, the simulation should have highlighted a single variant: the unsegmented-unsegmented version. This is not merely one version of the model; it is the prototypical organized anarchy.

## A World of Remarkable Order

Given this design, it is not surprising that the computer model leads to very odd and inappropriate implications for the behavior of organized anarchies, implications that clash with the informal theory. This is not obvious from the original article, which does not present detailed information on how the simulated organizations behave or what outcomes are produced. Instead, it gives summary statistics about things like "decision style" and "problem activity." But when the simulations are rerun (as we did, using the original Fortran program), the gulf between model and theory is strikingly clear.

This disconnect can be illustrated in many ways, for the underlying design problems leave their fingerprints almost everywhere. We will focus on one problem that is absolutely fundamental—and fatal.

The heart of the verbal theory is the notion of independent streams of participants, problems, and solutions intersecting unpredictably in organizational garbage cans—choice arenas—to generate organizational outcomes. This is basic. Perhaps the most important question we can ask of the computer model, then, is whether it reflects this central feature of the informal theory. The obvious place to look for this brand of chaotic behavior is in the paradigmatic organized anarchy, where both the access and decision structures are unsegmented. In these unsegmented-unsegmented organizations, anyone can work on any choice, any choice arena can be burdened by any problem, and virtually any permutation of choices, participants, and problems seems possible. If the model can produce garbage can–like behavior at all, one would surely expect to find it in this prototypical case.

To see, we reran the purely unsegmented version under all three energy loads: light, medium, and heavy. The results are illustrated in figures 9, 10, and 11. To simplify, we have scaled these figures down by half in every dimension: each depicts only five choice arenas, five decision makers, ten problems, and ten periods. This scaling loses no essential information and makes the figures easier to read. The text's descriptions of the simulation follow the figures and so are also scaled down by half.[12]

Consider the light load first (figure 9). In the initial period, one choice arena and two problems enter the simulation.[13] Because only one arena is open, both problems and all decision makers move—as a pack—to that one. The sum of the participants' energy inputs exceeds the problems' energy requirements, so the choice is made (via the energy-balancing mechanism described earlier), and the problems are resolved. In each of the next four periods, exactly the same thing happens: decision makers and problems travel in a pack from arena to arena; hence, each choice is made, and each problem resolved, immediately upon entering the simulation. The organization is extremely orderly and totally effective.

Under a medium load, with more energy needed to solve each problem, the organization does worse. Nonetheless, the main behavioral pattern is much the same: From period 3 on, all agents and all active problems move together in a pack (figure 10).[14] When a new choice enters the simulation after the second period, all players go to that arena. There they supply energy. But because this energy is less than the problems' energy

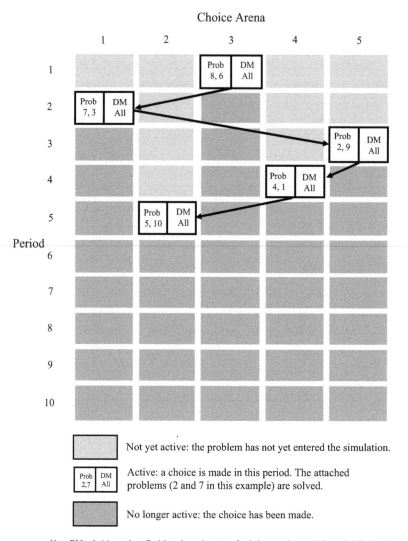

Figure 9. An unsegmented-unsegmented organization facing a light load.

requirements, the choice is not made. In the next period, the participants and problems all go to the next new garbage can that opens up. The old choice that they left behind is now "made," even though it contains neither decision makers nor problems. This happens because its ghost energy

Choice Arena

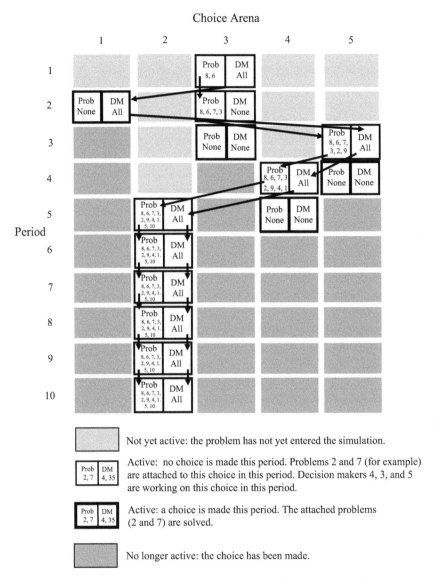

Figure 10. An unsegmented-unsegmented organization facing a medium load.

(energy cumulated over past periods) now exceeds the energy demanded of it—which is zero, simply because all the problems have left. In the second half of the simulation, no new choice arenas enter the process, so all the participants and problems must remain in one arena (the last one that entered) for the rest of the simulation. They have no place to go, because all the other choices have been "made" (with no problems or people present). In each period the decision makers exert more energy in that arena, but the simulation ends before enough has accumulated to solve the problems. Thus, the organization facing a medium energy load is completely ineffective. It solves no problems.

In a bizarre turnabout, however, going from a medium to a high load makes the organization totally effective again (figure 11). In the first period two problems enter and go to the one available arena. All participants go to this choice, again traveling in a pack, but they lack the energy to solve the problems. In the second period a new arena enters the simulation, and two new problems enter and go to this choice. The old problems remain at the old garbage can.[15] The entire pack of decision makers then goes to the new arena but again lacks the energy to solve the attached problems. In periods 3, 4, and 5, exactly the same thing happens: all the participants and two new problems go to the newly entered arena; all old problems stay at the old cans. At the end of period 5 all five arenas are active, each with two attached problems. Each choice has exactly the same energy balance because each has received one period's worth of energy from decision makers. In period 6, all decision makers return in a pack to choice arena 1. There they exert additional energy, make the choice, and resolve the attached problems. All participants then go en masse to arena 2, arena 3, and so on, in order. Hence, the pure unsegmented system facing a heavy load is completely effective, just like the organization facing the light load.

Thus, the three load versions present a peculiarly nonmonotonic pattern. Increasing the load initially makes the organization completely ineffective, but increasing the load once more sends problem-solving effectiveness soaring back to 100 percent. Sometimes nonmonotonic patterns are bold insights, but we doubt that this one is. The patterns bear no resemblance to expectations based on the informal theory. The latter certainly does not lead us to expect that a heavily burdened organized anarchy will be extremely effective at solving problems.

This is troubling, but the deeper problem is that the computer model generates clockwork dynamics that do not remotely resemble the decision processes associated with a disorderly garbage can. *The simulation's hallmark in the prototypical organized anarchy is that decision makers always move together in a single pack.* This pattern, moreover, is not confined to the unsegmented-unsegmented case highlighted here. It holds for all

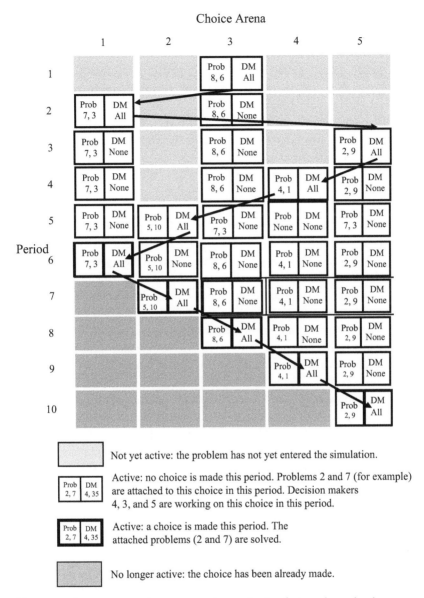

Figure 11. An unsegmented-unsegmented organization facing a heavy load.

simulations with unsegmented decision structures, regardless of the other parameters of the model. The reason is straightforward. In these simulations, all decision makers start in the same place (the first can that opens) and use the same rule to decide where to go in every period. These features, together with synchronous adjustment, imply that they virtually always stay together.[16]

It is difficult to imagine a more fundamental disjunction between model and theory. The computer model is supposed to represent the disorderly world of garbage can decision processes, but even in the prototypical case it generates an incredible degree of order. This is so clearly incompatible with the independent, randomly intersecting streams of the verbal theory that the simulations can offer no real insight into the workings of organized anarchies. They are from another world.

## A New Look at Well-Known Properties of the Garbage Can

Given the problems at work here, there is little point in examining the model's implications at length. They inevitably reflect the model's basic design, which is seriously at odds with the informal theory. We do, however, want to focus briefly on two well-known claims about organized anarchy that have emerged from these simulations. Over the years, they have been presented by the authors and others as basic properties of organized anarchies, and as central components of the theory itself.

The first is that "resolution of problems is not the most common style of making decisions" in organized anarchies (March and Olsen 1989, p. 13); decisions are more often made by flight or oversight. Decision making by flight and oversight are by now famous properties of the garbage can, conjuring up visions of a paradoxical organizational world in which many decisions get made, but problems are not resolved.

In fact, there is nothing paradoxical here. These properties are simply artifacts of the model's inappropriate assumptions. Flight occurs when all the problems attached to a choice arena leave, and the choice is subsequently "made." Oversight occurs when a choice is "made" before any problem attaches itself to the arena. In neither ease are decisions actually made; indeed, *nothing happens*. In ordinary English, decisions involve selecting an alternative. The model, however, uses its definitional fiat to brand truly vacuous events as "decisions." Because its own assumptions guarantee that empty choice arenas have zero energy requirements, its logic of energy balancing guarantees that all these "decisions" will be "made." The much-touted prominence of flight and oversight, therefore, is literally built into the model, an artifact of unwarranted assumptions; it says nothing about how organizations make decisions under ambiguity.

A second famous property associated with the garbage can is that decision makers and problems tend to track one another. This is presented as an interesting pattern in the otherwise mysteriously confusing complexity of garbage can dynamics. We make two points here. First, even if the model and its implications are taken at face value, the tracking phenomenon cannot properly be regarded as a general pattern of the simulation, for it does not and cannot hold for five of the nine variants. In systems with specialized access structures (versions 7, 8, and 9 in figure 8) problems can go to only one choice arena, so they cannot track participants. In systems with specialized decision structures (versions 3, 6, and 9) participants can go only to one choice arena; they cannot track problems. Only in the other four versions is it even theoretically possible for decision makers and problems to track one another. So claims that the tracking phenomenon is a central feature of the simulation would be misleading even if the model were acceptable.

But it is not. The model assumes (with a few minor wrinkles) that all decision makers and all problems make the same calculations simultaneously: they go to the arena with the smallest energy deficit. In general, therefore, when the access and decision structures allow them to go to the same, smallest-deficit arena, that is what they do. They go together, *in a pack*, which is another way of saying that they track one another. This, then, is our second (and more important) objection to the famous claim that tracking is a central property of organized anarchy: like the prominence of flight and oversight, tracking is simply guaranteed by the model's inappropriate assumptions about how participants and problems calculate where to go. It is an artifact of the design, and it tells us nothing about the dynamical patterns of organized anarchies in the informal theory, which clearly does not lead us to expect pack behavior.

Formalization often helps science progress, but it is not a panacea. In this case a heavy price was paid, for the verbal theory and the computer model do not represent the same phenomena at all. Decision makers in the verbal theory confront a chaotic world in which they, solutions, and problems dance around one another, meeting by chance in choice arenas. But in the simulation, packs of decision makers—and often problems as well—march in lockstep from arena to arena. And solutions never move at all.

This is ironic. The informal theory of the garbage can is famous for depicting a world that is much more complex than that described by classical theories of organizational choice. The latter's tidy image of goal specification, alternative generation, evaluation, and choice is replaced by a complex swirl of problems looking for solutions, solutions looking for problems, participants wandering around looking for work, and all three searching for choice opportunities. Yet, the simulation depicts almost none of this and in fact creates a world of remarkable order.

## Garbage Cans and Universities

The final section of the 1972 article applies the GCT to what the literature now routinely spotlights as the quintessential organized anarchy: universities. It aims to show that garbage can theory is more than just an abstract line of reasoning and can he used to shed new light on an important class of real organizations. This analysis has been influential (foreshadowing Cohen and March 1974 [1986] and March and Olsen 1976a) and is thus worth discussing here, if briefly.

The final section begins by asserting that universities conform to the basic implications of the theory, which it summarizes as follows:

> University decision making frequently does not resolve problems. Choices are often made by flight or oversight. University decision processes are sensitive to increases in load. Active decision makers and problems track one another through a series of choices without appreciable progress in solving problems. Important choices are not likely to solve problems. (1972, p. 11)

These implications are based without modification on the simulation, not on the verbal theory. This is an early, important, and very clear example of what has been going on in the literature ever since: the computer model and its implications have been bonded to the theory. The two have become one.

Applying the simulation is only the first step in this attempt to develop a garbage can analysis of universities. To derive more detailed predictions, the authors extend the model by adding auxiliary assumptions, which posit relations between the model's properties (e.g., access structure) and other variables, such as a university's size and wealth.

We do not object to the use of auxiliary assumptions per se. Introducing them is entirely legitimate, particularly when applying a general theory to a specific domain. But in this case the new assumptions are not part of GCT in any of its guises. They belong neither to the informal theory nor to the computer model. They are free-standing assertions, introduced casually and with relatively little discussion. Here are the sorts of claims they entail.

(1)   In bad times, the access structures of large rich schools are hierar-
      chical; those of small rich ones are unsegmented (Cohen, March,
      and Olsen 1972, p. 15, figure 4).

(2)   As times improve, the access structures of large rich schools change
      from hierarchical to specialized; the same environmental change has
      no effect on the access structures of small rich universities or poor
      ones (p. 15).

(3)   As times worsen, large rich schools and small poor ones change
      their decision structures; large poor schools and small rich ones do
      not (p. 15, figure 5).

No justification, theoretical or empirical, is given for these assumptions.
They are simply employed, along with the simulation, to generate predictions about university decision making, predictions that are supposed to
represent what GCT says about this important class of organizations.[17]

The predictions cannot do that, however, because they are based on
premises of sand. This is not a meaningful application of GCT to universities. The assumptions and implications of the simulation are largely
unwarranted and have little to do with the verbal theory. And the auxiliary
assumptions about universities are questionable, unrelated to the theory,
and offered with scant justification.

RECENT DEVELOPMENTS

We have now taken a close look at both key branches in the early development of GCT, the informal theory and the computer model. As our analysis
suggests, this early work left much to be desired. Yet such a story is not
unusual in the early stages of a research program. Pathbreaking work is
often crudely developed. Its real function is to shake up conventional ways
of thinking. With that done, the kinks, gaps, rough edges, and unexplored
implications can eventually be worked out to yield a coherent, well-defined
theory. But has this happened with the garbage can? Nearly three decades
have gone by, and a substantial literature has expanded upon the original
research. Let us see what this follow-on work has produced.

*The Computer Model as a Source of Confusion*

The computer model became a source of confusion almost immediately.
Although the 1972 article begins with a clear distinction between the "basic

ideas" and the attempt to model them via simulation, it later conflates the two. The GCT is summarized by statements such as "it is clear that the garbage can process does not resolve problems well," or that, quite often, decisions are made only "after problems have left a given choice arena or before they have discovered it (decisions by flight or oversight)" (Cohen, March, and Olsen 1972, p. 16). It is as though these simulation results, rather than being one set of implications from one possible model, were central to an understanding of garbage can processes. The model was fused to the theory.

In more recent work, this hybridized GCT, an undifferentiated blend of informal theory and computer model, is the norm rather than the exception.[18] Decisions by flight and oversight, access structures, decision structures, the tendency of decision makers and problems to track one another across choice arenas, the sensitivity of the system to variations in load—these and other components of the computer model tend to be presented as central, even defining features of garbage can processes in general, not as the highly specific assumptions or findings of one model.

This conflation of theory and model is especially debilitating because the computer model formalizes the theory so poorly. As the model's findings and building blocks have been tightly woven into the fabric of GCT, the theory itself has been transformed by claims and properties that are not legitimately part of the theory at all and that distort its real message. This would not have happened had the theory been granted a life of its own, separate from the model. Nor would it have happened had critics pounced on the model from the outset, exposed its inadequacies, and prevented it from gaining such prominence and influence. As things have developed, however, the garbage can literature has been profoundly and subtly shaped by an invalid model, to the point that GCT and its central ideas have become the model's illegitimate offspring. They are what they are, in good measure, because the computer model is what it is.

### The Computer Model, Science, and the Liberation from Science

We refer, here and elsewhere, to "the" computer model because Cohen, March, and Olsen did not develop other garbage can models after their 1972 article. A few others have tried their hand at it (Padgett 1980b; Anderson and Fischer 1986; Carley 1986).[19] Unfortunately, their models have not gained anything like the attention accorded the original, and their modifications have not been assimilated into mainstream organization theory's understanding of GCT.[20] Apart from this small body of technical work, the field has not benefited from a proliferation of formal models. The nontechnical literature on the garbage can has continued to center on just the original model, which has survived for nearly thirty years without challenge or change from the mainstream.

Throughout, this computer model has served as GCT's scientific core. However perplexing the authors' discussions of the garbage can, underpinning it all has been a literaturewide recognition that the informal ideas could be rigorously formalized, implications deduced, and tests conducted. The computer model was the justification, for it demonstrated, by sophisticated analytical means, that all this was indeed possible. It thereby bestowed scientific legitimacy on the entire GCT enterprise, no matter how unscientific it might otherwise seem. Thus, the computer model has enabled the rest of GCT to throw off the shackles of science and go its own way. The result has been an expanding, eclectic collection of loosely coupled ideas that have somehow seemed relevant to organized anarchy. Much of this is exceedingly complex and vaguely expressed, and how all these elements can be organized into a coherent and powerful theory is never explained.

## The Framework Expanded

Since 1972 the authors (primarily March and Olsen) have moved beyond their initial formulation to assemble a larger, more general theoretical apparatus. All of the more recent work is rooted in the GCT and has inherited, as a result, its fundamental problems. Indeed, in important respects the newer work is actually more complex and confusing than GCT ever was.

At the heart of this entire research tradition is a general theme that is a hallmark of the GCT's verbal theory: the juxtaposition of ambiguity and socially constructed order. The centrality of the first pillar was well reflected in the title and content of *Ambiguity and Choice in Organizations* (March and Olsen 1976a) and in many other publications.[21] But the second pillar is also central. Their work consistently emphasizes an interplay between the two. As March and Olsen (1984, p. 743) wrote in their seminal statement on the new institutionalism, "Institutional thinking emphasizes the part played by institutional structures in imposing elements of order on a potentially inchoate world."

In the 1972 article, ambiguity meant organized anarchy's triad of properties (ambiguous goals, unclear technologies, fluid participation) and socially constructed order—later called temporal order—that was allegedly produced by garbage can decision processes. This juxtaposition between latent chaos and institutional order is often referred to in later work.[22] But

new types of ambiguity and order have also been added. On the first dimension have been added the ambiguities of experience and the past (March
and Olsen 1975; Cohen and March 1986, pp. 199–201), of power and success
(Cohen and March 1986, pp. 197–99, 201–3), of relevance (March and Olsen
1976a, p. 26; March 1988, p. 390), of self-interest and deadlines (March and
Olsen 1976a, ch. 2 and p. 226, respectively), and of intelligence and meaning
(March 1988, pp. 391–95). On the second dimension the following have been
added: symbolic order (March and Olsen 1984, p. 744), normative order (p.
744), and, most prominently, interpretive order (March and Olsen 1976a,
chs. 4, 15; 1989, ch. 3).[23]

This sprawling theoretical framework is so complicated and ill specified
that there is very little chance of reformulating it in a coherent formal way.
Consider, for instance, its treatment of individual choice. March and Olsen
have often argued that their work is fundamentally about individual choice,
with deep roots in the bounded rationality tradition. Presumably because
the 1972 article did not actually use or develop any theory at the microlevel, a defining feature of their post-1972 work has been an effort to map
out a theory of individual choice that could be a microfoundation for the
new research program.

How has this been done? Although the authors claim they are engaging
in the "gradual relaxation of rigid assumptions in classical theories of
choice" (March and Olsen 1986, p. 28), and thus following in Simon's foot-
steps, *Ambiguity and Choice* actually abandons the framework of bounded
rationality—and of rational choice, too—and strikes out on its own. The
result is a free-floating discussion of decision making that has no clear
framework. Here are some of the factors that the book argues should be
included in a theory of individual choice.

(1)  Such a theory must recognize that "beliefs and preferences appear
     to be the results of behavior as much as they are the determinants of
     it. Motives and intentions are discovered post factum" (March and
     Olsen 1976c, p. 15). Hence, the bedrock components of decision
     theories are to be treated as endogenous. The problem is that,
     although exogenous parameters are vital to the generation of pre-
     dictions in any testable theory, nothing about individual choice is
     taken as exogenous here.

(2)  It must recognize that individual preferences are shaped less by self-
     interest or problem solving than by roles, duties, and obligations,
     and by "the definition of truth and virtue in the organization,
     the allocation of status, the maintenance or change of friendship,

goodwill, loyalty, and legitimacy; [and] the definition and redefinition of 'group interest' " (p. 16).

(3)   It must recognize that beliefs are shaped by decision makers' interpretations and by "myths, fictions, legends, and illusions" (p. 18). All this and more would be part of a larger theory of learning that is needed to explain the evolution of beliefs.

(4)   It must recognize that people do not attend to all issues. Their involvement is selective, often for reasons unrelated to substance, and any explanation of this behavior—a theory of attention—must emphasize "duty, tradition, and routine"; the "educational, ideological, and symbolic role of choice situations in organizations"; and the fact that each person's opportunities depend on what everyone else is doing (March and Olsen 1976b, p. 45).

This book's perspective on choice, which was complicated even further in later work, may be a plausible description of the countless forces that affect people's decisions. But description is not theory. And we do not see how this enormously complicated tangle of possibly relevant factors can ever lead to a rigorous, productive theory. Theories are supposed to reduce complexity, not surrender to it.

Despite these difficulties, *Ambiguity and Choice* quickly had a substantial effect on the field of organizations. Its influence on political science lagged a bit, but in the 1980s several works that were highly visible to political scientists disseminated its ideas in a major way. First, the *American Political Science Review* published "What Administrative Reorganization Tells Us about Governing" (March and Olsen 1983), in which garbage can concepts figured prominently.[24] The research program got an even bigger boost the following year when the journal published "The New Institutionalism" (March and Olsen 1984). This much-cited work followed *Ambiguity and Choice* in "de-emphasiz[ing] metaphors of choice and allocative outcomes in favor of other logics of action and the centrality of meaning and symbolic action" (1984, p. 738). Furthermore, among the article's six "institutional conceptions of order," three are clearly linked to the 1976 book or the 1972 paper: temporal order (pure GCT), normative order, and symbolic order.[25] And when the article suggests three "examples of possible theoretical research" (1984, pp. 744–46), it is no accident that one of these is the garbage can model.

The 1984 article, then, brought substantial attention within the discipline to March and Olsen's variant of institutional thinking, which they

presented as an integral part of the new institutionalism. What solidified
its status, however, was the more elaborate presentation of these ideas
in *Rediscovering Institutions* (March and Olsen 1989). The book was
quickly hailed as a major contribution and, as noted above, has come
to be regarded as a contemporary classic (Goodin and Klingemann 1996,
p. 16).

Obviously, we cannot analyze this complex book in detail. But its
anchoring in earlier GCT work—especially *Ambiguity and Choice*—is
evident throughout, and we will simply offer a few (of many possible)
illustrations of its intellectual lineage.

(1)   Chapter 3, "Interpretation and the Institutionalization of Mean-
      ing," begins with a section on how people make sense of their
      world. This was a central concern of *Ambiguity and Choice*. But the
      similarity goes far beyond common themes: Much of this section in
      the new book (pp. 41–45) is taken word for word from the earlier
      book.

(2)   One of the most-noted themes in *Rediscovering Institutions* is the
      contrast between "the logic of appropriateness associated with
      obligatory action and the logic of consequentiality associated with
      anticipatory choice" (1989, p. 23). This same theme was developed
      in *Ambiguity and Choice*, in its analysis of "attention as obligation"
      (pp. 48–50).

(3)   Most of the chapter on institutional reform is based on March and
      Olsen's 1983 article, which in turn essentially restates the argu-
      ment developed years earlier in Olsen's chapter, "Reorganizations
      as Garbage Cans," in *Ambiguity and Choice*.

(4)   The references section of the 1989 book gives individual citations
      to all but one of the seventeen chapters in the 1976 book, and
      *Ambiguity and Choice* is far and away the most-cited item of their
      references.

In short, the GCT of the 1972 article led to the interpretivist vision of
the 1976 book, which in turn led to the new institutionalism of the 1984
article and the 1989 volume. There has been growth and elaboration (if
not progress) throughout this evolution, but the continuity in themes and
substance is striking. These writings clearly constitute a single research
program, which carries the genes of the garbage can (Sjoblom 1993).

This research program does have intriguing ideas and assertions, but
the problems remain and are much the same. The formulations are overly

complex, the arguments unclear. Consider one of the best-known themes of *Rediscovering Institutions*, the importance of the "logic of appropriateness" in politics.[26] The core hypothesis is that "most behavior follows such a logic of appropriateness, that rules are followed and roles are fulfilled" (p. 161). This may seem to be a straightforward and empirically falsifiable claim about politics, but in fact it is neither. Whatever the logic of appropriateness refers to, it clearly involves rule-governed behavior. To know what it implies and how it might be tested, then, we must know what rules are. Early in the book the authors deal with this explicitly. What they do, however, is take a concept with clear meaning in ordinary language and transform it into a conceptual morass: "By rules we mean the routines, procedures, conventions, roles, strategies, organizational forms, and technologies around which political activity is constructed. We also mean the beliefs, paradigms, codes, cultures, and knowledge that surround, elaborate, and contradict those roles and routines" (p. 22).[27] This staggering expansion of the concept trivializes the claim that political behavior is rule governed. Since the authors' definition leaves so little out, what else *could* drive institutions? This expanded definition also makes it impossible to figure out what the logic of appropriateness refers to in concrete terms, and thus what kinds of behaviors are consistent or inconsistent with it. There is no clear content here to test.

In addition, *Rediscovering Institutions* never makes clear how rule-driven behavior differs from instrumental behavior. The predominance of the former and the relative insignificance of the latter is one of the book's central theoretical claims; yet at times the authors seem to recognize that rule following can be instrumentally based and that the two are not at all distinct.

> To say that behavior is governed by rules is not to say that it is either trivial or unreasoned. Rule-bound behavior is, or can be, carefully considered. Rules can reflect subtle lessons of cumulative experience, and the process by which appropriate rules are determined is a process involving high levels of human intelligence, discourse, and deliberation. Intelligent, thoughtful political behavior, like other behavior, can be described in terms of duties, obligations, roles, and rules. (p. 22)

Any notion that there is a clear distinction between rule following and instrumentalism, moreover, is contradicted by cognitive science. The logic of appropriateness involves matching actions to situations (p. 23)—as in "if in situation x, do y"—but cognitive scientists argue that this is a broad rubric and very often involves rules that are cognitive guides for instrumental action, encoding or summarizing the problem-solving effectiveness of skilled (although boundedly rational) agents (Anderson 1995). In a larger social setting, some of these instrumentally based rules, such as standards of professionalism, may emerge as social norms.[28]

The confusion surrounding the logic of appropriateness could have been dispelled had the authors anchored their variant of the new institutionalism in a coherent microtheory. Ironically, one was right at hand: March and Simon (1958, p. 139–50) analyzed how cognitively limited decision makers would be guided by rules and routines, or what the authors called performance programs. There is now a substantial literature, some of it quite formally sophisticated, that seeks to explain organizational behavior on these individualistic foundations. Using the same foundations would have encouraged March and Olsen to examine individual choice in a clear, rigorous way and would have given them a framework of demonstrated analytic power that anchors organizations in the choices of individuals. It also would have allowed them to generate detailed and testable implications, thus encouraging the kind of empirical research and theoretical adjustment to evidence that characterizes all scientifically progressive fields.

Why didn't the authors ground their version of the new institutionalism on the Simon-March research program? Their framework cries out for a microfoundation that the latter could easily provide—yet they do not use it. Given their intellectual sympathies for the Simon-March tradition, this is puzzling. The answer may turn on a self-inflicted wound: their stance against reductionism (March and Olsen 1984, pp. 735–36, 738; 1989, pp. 4–5). They seem to view this position as a key feature of their new institutionalism. If one accepts it as a metapremise (don't build theories of institutional behavior on individualistic premises), then the option of building their new institutionalism on the bounded rationality program is simply precluded.[29] But this merely pushes the puzzle back another step: why is antireductionism a central tenet of their approach? Why did they constrain their theory construction in this manner? We do not know. But if the reasons for their methodological choices are not apparent, the effects are. Their arguments tend to be unclear and much too complicated, due in large measure to the absence of a disciplined microfoundation.

## Chaos, Order, and Causality

Work on GCT has been much cited in organization theory, political science, and sociology. Yet, apart from work done by the authors themselves and some of their colleagues, the garbage can and its descendants have been put to little empirical use in these fields. For a theory often considered one of the major perspectives on organizations (e.g., Perrow 1986; Scott 1992), this neglect might seem odd, but we do not find it so. As it stands, the theory is almost impossible to test, and this can only discourage serious empirical research.

Many applications of the approach do little more than describe some parts of an organization as garbage cans, offering descriptions that emphasize GCT's central themes.[30] These case studies are, as Anderson and Fischer (1986, p. 141) note, often "only loosely coupled to the theory" (for a similar criticism, see Perrow 1977, p. 297). They describe events using garbage can terminology, but it is not at all clear that they provide real GCT *explanations*. Authors of such applications (e.g., Sproull, Weiner, and Wolf 1978) tend not to see it this way; they seem to believe that, by merely labeling and describing organizations along these lines, they have somehow helped us understand them—that description has produced explanation. But it hasn't.

Latent here is a methodological issue of great importance: what counts as an explanation, and what kind of explanation do GCT scholars aim to provide? Clearly, description is not explanation; to equate the two is a mistake. But what is the answer to the larger methodological question? If GCT and its offshoots did their job properly, what kind of explanations would they provide? How would they help us understand organizations?

Addressing this issue is crucial if the authors are to put their work in scientific perspective. And in fact they do address it, although in less depth than it deserves. They argue that their type of explanation is methodologically very different from those of conventional theories: it rejects the "consequential order" of conventional theories, substituting a "temporal order" in which "problems, solutions, and participants are joined together more by the timing of their arrivals than by other attributes" (March and Olsen 1986, p. 12).

Here, the authors emphasize that the theory is not only about chaos and disorder, the feature that has caught the attention of most social scientists, but also about order. Through the lenses of conventional theories, organized anarchies look chaotic, but they have an underlying order that the theory can help us see. This is good to point out, especially given how others have interpreted their work. But they still need to tell us how temporal and consequential orders differ, and in what sense the two give different types of explanations. So far, these issues remain unclarified.

We are reasonably sure what the authors mean by a "temporal ordering": events are linked—ordered—based on the timing of the various streams' intersections. We are less sure what "consequential ordering" means. It seems to be a perspective that explains organizational outcomes via the intentions of individuals. It is "consequential," presumably, because actors intentionally make choices—and thus generate organizational outcomes—by assessing consequences.

From this vantage point the authors' methodological claim seems straightforward: GCT explains organizational outcomes on the basis of timing rather than intentions. Yet if this is all there is to the argument, no new language is needed; and without the new language there would be no confusion. If we take the new concepts and the accompanying (brief) discussions seriously, however, there is much to be confused about and much to pay serious attention to—for it appears that what the authors are really doing is distancing the garbage can framework from causal explanations in general.

This is reflected in a second, more general way of thinking about consequential order that sometimes seems to orient the authors' analysis. In this interpretation any causal sequence involves a consequential logic—that is, certain outcomes are effects of certain determinants. X causes Y; hence Y is a consequence of X. Intentionality rests on such a logic, but this does not distinguish it from other theories. All causal theories use a consequential logic, in this sense.

Had the authors meant *only* intentionality when using the term *consequential order*, as in the first interpretation, then presumably they would have said so explicitly. But they didn't. We suspect this is because they were referring to more than intentionality—directing their criticism, at least implicitly, to causal approaches in general.

This inference is reinforced by the way the authors discuss temporal ordering. The essence of a temporal ordering, in their portrayal, is not just that timing determines outcomes; it is that events are not driven by an identifiable causal structure. Never, in fact, do Cohen, March, and Olsen present such a causal structure as the foundation of their theory, and they rarely say that identifying and modeling it are the keys to explanation, as they would be in conventional theories.[31] In their view, the flow and intersections of events are largely driven by random occurrences, accidents, and a huge variety of complex institutional, social, psychological, economic, and political forces that cannot be represented by a causal model. The upshot is a perspective in which outcomes are explained not by a well-specified causal structure but by the timing of events and their random, unspecified determinants. As a statistician might say, all the action is in the error term.

Because the authors' methodological discussions are so brief, we cannot be sure that we have correctly described their position. Yet, in a more recent book March seems to confirm our judgment, for he explicitly states that his framework entails "the orchestration of decisions through temporal orders *rather than causal orders*" (1994, p. 180, emphasis added). In any event, if GCT and its extensions are ever to be legitimate scientific theories, they must set out a causal logic of some kind. That timing is central to their explanations has no bearing on the causality issue. A proper dynamic theory differs from other theories only in having a time component; all theories worthy of the name, including dynamic ones and those allowing for randomness, identify some causal structure as the basis for explanation. To suggest that their temporal order is relieved of this responsibility, that it can explain events without delineating causation, creates a fundamental confusion.

Finally, this methodological confusion ties into the garbage can's empirical applications in an ironic and unproductive way. It is hard to read this literature without encountering a convoluted line of reasoning that goes roughly as follows. What we do not understand about certain organizations—because their behaviors appear horribly complex, seemingly random, and impossible to disentangle—is actually just what the garbage can theory tells us to expect. In this sense we do understand it. Although we have almost no idea why certain events occur, we "understand" them because they are precisely the kind of incomprehensible phenomena that garbage can processes produce. That organizations are so confusing, then, *is actually strong evidence of the theory's great explanatory power.* The less we can figure anything out—the less, that is, we can identify a causal structure—the more the theory seems to be working, and the better a theory it seems to be.[32]

This reasoning may seem far-fetched, but anyone familiar with the literature knows it is quite common. The solution is simple: GCT, if it is to become a genuine theory, must play by the same rules as do all other theories in social science. The job is to identify the causal structure of institutional behavior. That is how behavior is explained and understood.

## REVITALIZING THE GCT

Despite the many problems we've identified, we believe that the garbage can program can be revitalized. Because we cannot carry out a full reconstruction here, we merely provide an example to show that basic themes of the GCT can be grounded in theories of bounded rationality. Our example will show how the well-known property of temporal ordering—that solutions may be linked to problems more by chance than design—can be derived as a natural implication of a classical model of adaptive search (Simon 1957). This is a significant test case because temporal ordering looms large in the GCT literature.[33]

We begin with Simon's satisficing model and assume that there is an organizational superior who decides whether to accept an alternative to solve a particular problem. This manager has three subordinates, trained in different professions. (Imagine a diplomat, a military officer, and an economist working for the secretary of state.) Due to the trained incapacity of specialists, the diplomat sees the problem as one of diplomacy, the officer sees a military problem, and the economist sees an economic one. They craft their solutions accordingly. The quality of proposals and how long it takes to generate them are random variables. Once a proposal is done, it goes to the boss. He or she has an aspiration level that acts as a stopping rule: if exactly one proposal exceeds the boss's aspiration, he or she selects that one and the process ends. If no proposal satisfices, the staffers keep working. If more than one does, the boss picks the one he or she thinks is best.[34] Thus, the superior picks samples from three different distributions, produced by the three specialists.

The value of the superior's aspiration level powerfully influences this process, as the following (easily proven) implications show. (1) If the boss's aspiration level is very demanding, then the randomness of proposal generation will have little effect on which type of solution is accepted (although it will affect how long it takes to find a satisfactory option). If his or her aspiration level is so high that only one type of proposal can satisfy it, then random generation has no qualitative effect on the outcome. (2) *The lower the superior's aspiration level, the more the process exhibits temporal coupling*—that is, the more the proposals' chance arrival times influence what type of solution is picked. If the boss's aspiration is below the worst possible alternative, then choice is driven completely by chance.

This shows how a central GCT property can easily be generated by a classical model of behavioral search. But it also suggests how the prediction of temporal coupling may be empirically limited.[35] The preceding sketch fixed aspirations exogenously, as they were in Simon's model. But many have argued that aspirations adjust to experience, rising in good times and falling in bad (e.g., Cyert and March 1963; Bendor, Kumar, and Siegel 2004). If aspirations were thus endogenized, then temporal coupling would tend to diminish over time. The superior, discovering that the payoffs beat his or her initial standard, would tighten his or her requirements and therefore become less prone to accept poor solutions.

Temporal coupling has been held up by garbage can theorists as a finding of almost mystical significance. But it turns out to be very compatible with choice-theoretic formulations. So, we believe, are all the other GCT ideas that are worth keeping.

## CONCLUSION

The philosopher Max Black (1962, p. 242) once remarked that science moves from metaphor to model. In part he meant that science *should* so move. As he went on to say, metaphors are invaluable at the start of an inquiry. Ideas often come to us first as metaphors, a vague notion that *x* is like *y*. These can be creative insights; they may revolutionize a field. But they should not remain as they are born: as scientific formulations, metaphors are flawed (Landau 1972). Their logic is obscure. How do claims relate to each other? Which are premises and which conclusions? And the empirical content of metaphors is often thin: few, if any, of their claims may be empirically testable. Thus, both to clarify their reasoning and to provide targets for empirical scrutiny, scientists should—and often do—transform metaphors into models.

The garbage can research program has done just the opposite: it has moved from model to metaphor. A model was presented first, in the same pathbreaking article that presented the metaphorically inspired informal theory. Later, instead of critiquing and revising the original formal model—which certainly needed attention—the research program focused mainly on embellishing the verbal theory. And it did so in ways that not only failed to clarify GCT but actually made it *more* metaphorical.

Because the initial computer model misrepresented the informal theory, perhaps tinkering with it would have been a mistake. It probably should have been replaced by a new model. This did not happen. Instead, the original simulation was frozen in time.

The fault should not be laid primarily at the feet of the authors. Their 1972 article included a copy of the Fortran program; everyone had a chance to critique it. The article's text described the simulation well enough so that the discrepancy between the informal theory and formal model could be discerned. So the authors discharged their obligations by giving readers plenty to work with. By and large, readers passed up this opportunity. Thus, the verbal theory and the simulation remained so deeply conflated that few could tell where one left off and the other began. That is no way to create a research program with ever-clearer logic and growing empirical content.

We believe it is possible to revitalize the GCT. Doing so requires radical surgery. Because this operation would make the framework's logic transparent to all, it would deprive the GCT and the March-Olsen variant of the new institutionalism of a certain mystique. Without this bold move, however, there is little chance that these ideas will shed much enduring light on institutions.

## POSTSCRIPT: A MODEL OF SOME GARBAGE CAN PROCESSES

Bendor, Moe, and Shotts (2001) argued that it is possible to revitalize the GCT.[36] Here we present one such attempt to capture some major aspects of garbage can processes that were stipulated by the original verbal theory. Our model focuses especially on problematic preferences. We consider an institution headed by a committee of three decision makers. Each person has an ideal policy in a two-dimensional policy space and preferences are *spatial*, in the sense used by political scientists: the farther a policy is from person $i$'s ideal point, the less he or she likes it. Hence, indifference curves in this two-dimensional space are circles. The decision makers' ideal points are equidistant from the center of the policy space, which is normalized to $(0, 0)$.

Solutions are new policies. In every period, a new policy is drawn from a uniform distribution on the circle shown in figure 12.[37] The committee members don't understand how new policies are produced (unclear technology): they simply compare, in any given period, the new policy to the status quo and adopt whichever one is preferred by a majority. For example, in figure 13 the new policy, $x_2$, is closer to the ideal points of decision makers 1 and 2 than is the status quo ($x_1$), so $x_2$ beats $x_1$ and becomes the new status quo.

Thus, the decision makers are myopic—focussing only on the present vote and ignoring what might happen in the future—and in this sense are boundedly rational. (For a model of the behavior of farsighted, fully rational decision makers in a similar setting, see Penn 2007.)

The institution's preferences are problematic in several ways. First, there is no policy that beats all others in pairwise votes: an example of Condorcet's famous problem. Second, and worse, the organization's preferences are completely intransitive: McKelvey's theorem (1976) implies that for every pair of policies $x_1$ and $x_n$, there is a set of policies $\{x_2, \ldots, x_{n-1}\}$ such that $x_2$ is majority preferred to $x_1$, $x_3$ is majority preferred to $x_2$, and so on all the way to $x_n$, which is majority preferred to $x_{n-1}$—yet $x_1$ is majority

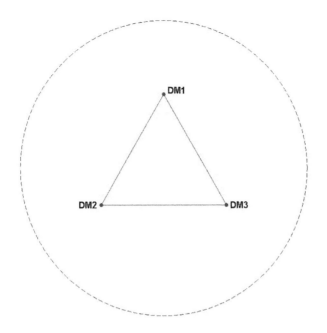

Figure 12. The committee and its policy space. DM = decision maker.

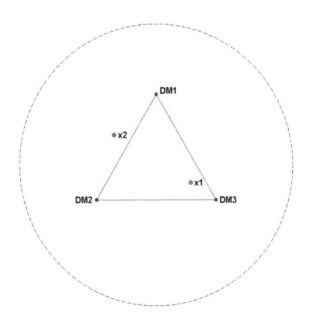

Figure 13. Simple myopic voting. Under simple myopic voting, x2 beats x1.

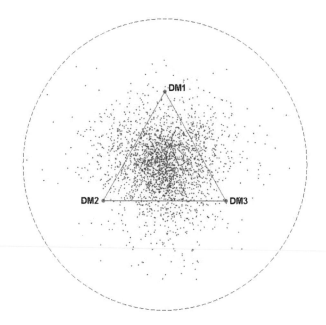

Figure 14. The process with fixed participation.

preferred to $x_n$. In a fundamental sense, the institution's preferences are incoherent—problematic indeed!

First we present a version of the model with fixed participation: all three committee members are present in every period. Once we understand this model's implications, we will introduce fluid participation.

### The Model with Fixed Participation

We simulated this process 2,500 times, letting each process run 1,000 periods.[38] It is easy to show that the associated stochastic process—a finite Markov chain with stationary transition probabilities—is ergodic: it has a unique limiting distribution and the process converges to this unique distribution from any starting point.[39]

Figure 14 depicts a scatter-plot of the status quo policy in each of the 2,500 runs in period 1,000. Most of the policies are centrally located; a three-dimensional plot would show a a distribution that is single peaked and centered at (0, 0).[40] In comparison, recall that new alternatives are drawn from a distribution that is smeared uniformly over the entire policy space (i.e., the circle).

Thus, even though organizational preferences are problematic and the generation of alternatives is exceedingly random—any feasible policy is as likely to be dumped into the in-basket as any other—the process's outcomes exhibit a clear pattern: they tend to be centrally located. True, they are still stochastic—nothing is *bound* to happen—but some things are more likely than others. And we can understand why: although the decision makers do not understand their own institution's technology and although they focus myopically on the here and now, what they want still matters. Status quo policies that are far away from the ideal points of the entire committee are easily and often beaten. Those that are close to the ideal points are hard to beat. So the latter, though not perfectly stable, are much more stable than the former. (For an insightful early statement of this argument, see Ferejohn, Fiorina, and Packel 1980 and Ferejohn, McKelvey, and Packel 1984.)

## The Model with Fluid Participation

In this version each decision maker shows up in every period with probability $\frac{1}{2}$.[41] Each person's coming to a meeting is independent of the behavior of the other committee members. Therefore, the chance that all three participate in a given period is $(\frac{1}{2})^3 = \frac{1}{8}$, the chance that two participate is $\frac{3}{8}$ (three different possible pairs, all with $\frac{1}{8}$ chance), and so on. Consistent with standard views about organizational inertia, the status quo policy is privileged: if nobody shows up at $t$, then the status quo policy carries over into $t + 1$; if two decision makers arrive and they split over the merits of the new alternative, then again the status quo prevails.

Figure 15 gives a scatter plot of the status quo policy in period 1,000 (again, for 2,500 runs). The most striking effect of fluid participation is that outcomes become more spread out, as we can see by comparing figure 15 to figure 14. This happens because often only one decision maker comes to a meeting and at such times he or she is dictator for the day: if this person prefers the new policy to the status quo, then the former triumphs. Further, with a single decision maker it is possible to jump far from a centrist status quo: for example, the process can go in one jump all the way to the lone decision maker's ideal point—or even well beyond it. This cannot happen if all three people are present. With fixed participation in a majority-rule institution, the status quo is beaten only if more than half of the decision makers prefer that. Hence, it's easier to overturn centrist status quo policies when participation is fluid then when it is fixed, and fluid participation also permits bigger departures from the center.

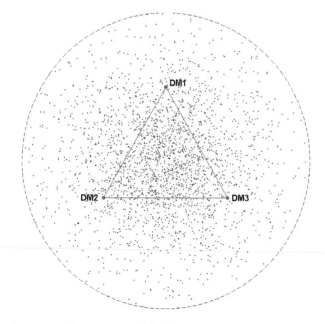

Figure 15. The process with fluid participation.

We conjecture that if participation is semifluid—for example, each decision maker shows up on any given date with probability 0.75—then the variance will be between completely fluid participation (the 0.5 case) and the fixed participation case.[42]

Thus, garbage can processes can be rigorously analyzed. They produce understandable results which, though stochastic, aren't fundamentally mysterious.

## CHAPTER 7

# Institutions and Individuals

JONATHAN BENDOR

Everyone—leaders as well as citizens—is cognitively constrained. Psychologists have been demonstrating this in carefully controlled experiments for decades. To take one prominent example, the field of human memory has examined this theme intensively. (For an excellent introduction to what is known about memory errors, see Schachter 2002.) So the only serious questions for political science and public policy analysis are how these constraints affect variables that concern us directly: political behavior, the utility that citizens get from policies, and so forth. On this point, evidence has been growing for several decades (e.g., Jervis 1976; Kinder 1998; Tetlock 2005) that cognitive constraints affect variables of interest to political scientists and policy analysts.[1]

But the cognitive limits of individual decision makers is only the beginning of the story. Stable polities are governed at least as much by institutions as by individuals, so the bounded rationality (BR) program must analyze information-processing constraints at both levels. In this concluding chapter, I pursue a specific aspect of this topic that relates both to theoretical political science and to practical public policy: how institutions can compensate for the bounded rationality of individuals. This issue was near to the hearts of both Aaron Wildavsky and Marty Landau, though in somewhat different ways: Aaron emphasized professional training (hence, the Graduate School of Public Policy), while Marty focused on structural approaches (especially redundancy). They transmitted this interest to many of their students, present company included (Bendor 1985). So ending with this topic is fitting.

## THE CONVENTIONAL AMERICAN VIEW: PLODDING
## BUREAUCRACY VERSUS THE HEROIC INDIVIDUAL

It is widely believed that the United States has an individualistically oriented culture. It turns out that this is more than a belief: systematic research (Triandis 1995) indicates that it's a fact—the United States is one of the most individualistic cultures in the world. This has important implications for this chapter's topic. It is no accident that a popular theme in entertainment and the media is how a heroic and determined individual gets something done despite obstacles thrown in his or her path by a plodding or mindless bureaucracy. A careful content analysis of popular culture would likely show that the opposite story—how an agency collectively solved a problem despite the errors of fallible individuals—is much less common.

This pattern may also hold in academia. A startling example is provided by Graham Allison's summary and use of the Simon-March-Cyert line of work in his famous *Essence of Decision*.[2] I focus on Allison because of the impact of his article (1969) and his book (1971): the former is the most downloaded article of the *American Political Science Review* in almost a century (1906–2003); the latter has been on several short lists of influential books in the discipline (e.g., Goodin and Klingemann 1996, p. 16). And for political scientists, Allison's model II is probably the most-used summary of the Carnegie tradition. So its interpretation of that body of work has been consequential.

Consider the following passages: "Simon and the Carnegie School focus on the bounded character of human capabilities. Firms are physically unable to possess full information, generate all alternatives" (Allison 1971, p. 174); "The physical and psychological limits of man's capacity as alternative generator, information processor, and problem solver constrain the decision-making processes of individuals and organizations" (p. 71). These passages move very swiftly from individual constraints to organizational ones. But when we turn to Simon's seminal book, *Administrative Behavior*, we see that Professor Allison *got it exactly backwards*. In Simon's view, organizations typically compensate for the bounded rationality of individual decision makers:

> It is impossible for the behavior of a single, isolated individual to reach any high degree of rationality. The number of alternatives he must explore is so great, the information he would need to evaluate them so vast that even an approximation

to objective rationality is hard to conceive .... One function that organization performs is to place the organization members in a psychological environment that will adapt their decisions to the organization objectives, and will provide them with the information needed to make these decisions correctly .... In the course of this discussion it will begin to appear *that organization permits the individual to approach reasonably near to objective rationality.* (pp. 79–80, emphasis added)

This is not an attempt at proof by authority. Simon was brilliant, but his hypotheses could be wrong anyway. The point is that in chapters 3 and 4 of his book, Allison aimed to summarize the Carnegie School and to use its ideas to analyze the Cuban missile crisis. In this context, the gap between the original formulation and Allison's interpretation is meaningful.[3] This is, I believe, a paradigmatic example of how an unusual idea is mentally assimilated to an older, familiar one. And since the new idea challenged conventional wisdom, it could have been valuable to the discipline. Thus, because an influential article and book misinterpreted Simon's unusual hypothesis, our collective mental repertoire was impoverished.

## INSTITUTIONAL APPROACHES TO DEALING WITH THE COGNITIVE CONSTRAINTS OF INDIVIDUALS

Let us now briefly examine some ways in which institutions compensate for the cognitive constraints of individuals. First we'll look at phenomena that can be observed in the short run: days, weeks, months. Then we'll stand back and examine a long-run process: modernization.[4]

### Short-Term Processes

*From Serial to Parallel Processing*    In a neglected passage of *Organizations,* March and Simon note an important difference between individual and collective cognition: "The individual possesses only a single focus of attention, hence can only deal with one aspect of the problem at a time. In organizations, however, there is no limit to the number of attention centers that can be allocated to the parts of a problem" (1958, p. 193). They then note the advantages of parallel processing. Clearly, if one large problem can be decomposed into many subproblems, an agency can greatly speed up its information processing by assigning a different official to each subproblem.[5] Of course, there is no guarantee that any given division of labor will be an effective decomposition. The trick, as many students of organizational design have noted, is to figure out a scheme of specialization that carves nature at the joints (Gulick 1937; Simon 1947; Simon, Smithburg, and Thompson 1950; March and Simon 1958). When this can be done, the serial constraint can be substantially eased.

This organizational advantage is obvious. Just mentioning the phrase *division of labor* helps us recognize an organization's advantage over an individual. In the simplest model of problem decomposition (an additive one) information processing is essentially the same as performing a physical task, like Frederick Taylor's (1947) pig iron loading. Obviously, the more people, the more pig iron is loaded or the faster information is processed. Otherwise, why bother to hire the additional people? Yet in the organizational decision theory literature, this obvious point is sometimes overlooked.

*The Division of Labor*     It's not just that work in complex organizations is done in parallel. People specialize, and as social scientists since Adam Smith have argued, specialization facilitates learning. The mechanism is simple: it's learning by doing. The more often one does a task, the faster one goes and the fewer errors one makes (Argote and Epple 1990). This phenomenon of the *learning curve* is one of the strongest empirical regularities of performance that we know about. We obtain this effect by dividing up tasks so that individuals can do the same task over and over rather than jump around to different tasks. The effect, in domain after domain, is learning by doing. Moreover, it holds at organizational levels as well, whether the task is building airplanes (Argote and Epple 1990) or performing medical procedures (Gawande 2002).

*Routines*     Routines are much despised in the conventional wisdom; they are part and parcel of the conventional negative stereotype of bureaucracies as clunky entities mindlessly executing procedures.[6] This pattern was reinforced by Allison's *Essence of Decision*. Scholars typically understand the main lesson of model II to be that routines and standardized scenarios constrain and rigidify organizational behavior (e.g., Krasner 1972, pp. 169–75; Art 1973, pp. 476–79; Gallucci 1975, p. 145; Peterson 1976, p. 113; Jefferies 1977, pp. 231–32; Williamson 1979, p. 140; Thompson 1980, p. 27; Scott 1981, p. 6; Levy 1986). (For a partial exception to this pattern, see Posen 1984, pp. 46–47.) True, Allison occasionally mentions the possibility of organizational innovation and at times portrays standard operating procedures as efficient ways of handling standard problems.[7] But most of chapter 3's theoretical exposition emphasizes the negative, constraining effects of organizational routines. And this pattern is strikingly reinforced in chapter 4's empirical application of model II to the missile crisis. Virtually without exception, the episodes covered there describe how the permanent bureaucracies of both countries fouled things up. Toward the end of chapter 4, Allison pointedly asks, "Were the organizations on top of which the President was trying to sit going to drag the country over the nuclear cliff in spite of all his efforts?" (1971, p. 141). The heroes of this chapter are clearly the unbureaucratic president and his advisors. Therefore, the common interpretation that model II's main lesson is the rigidifying, maladaptive effects of bureaucracy is well founded.

This is an irony of intellectual history. As Allison noted, the central ideas of model II are rooted in the Carnegie School. Yet that tradition's founding volume, Simon's *Administrative Behavior,* views routines far more positively:

> Habit performs an extremely important task in purposive behavior, for it permits similar stimuli or situations to be met with similar responses or reactions, without the need for a conscious rethinking of the decision to bring about the proper action. Habit permits attention to be devoted to the novel aspects of a situation requiring decision. A large part of the training that goes to make a championship football team, crew, army battalion, or fire company is devoted to developing habitual responses that will permit immediate reactions to rapidly changing situations.
>
> Habit, like memory, has an artificial organization counterpart, which has been termed by Stene "organization routine." (1947, p. 88)

Hence, chapter 4 of *Essence of Decision* inverts Simon's view. Instead of revealing how organizations can boost individual rationality, chapter 4 argues that the bureaucracy's rigid conduct nearly nullified President Kennedy's best efforts.[8] In contrast, Simon saw routines as encoding organizational memory. Hence, the buildup of routines often reflects organizational learning. As evolutionary anthropologists (e.g., Richerson and Boyd 2005) have argued, it is this cumulative learning—that is, culture—that explains our explosive takeover of the planet. Individuals lack the time and cognitive bandwith to learn most of what they need to know by themselves.

*Reliable Systems, Unreliable Components*    In his book on the Polaris missile system, Sapolsky (1972) observed that several teams worked independently to develop a key component. The U.S. Navy's Special Projects Office deliberately established this redundant structure to increase the odds that the essential component would be ready by a specified date. By doing so, the Special Projects Office loosened the connection between the reliability of an organizational subunit (one of the problem-solving teams) and the larger organization—in contrast to the message of Allison's model II.

More generally, a properly designed agency can display large disparities between the reliability of different organizational levels (Landau 1969; Grofman, Owen, and Feld 1983; Bendor 1985). Consider a representation of the Polaris project. There are $n$ teams trying to develop the same component. The probability that any one of the teams will succeed in the specified time is $p$, where $0 < p < 1$. If the performance of the teams is independent, the probability that the organization will succeed equals the probability that at least one of the teams succeeds, which equals $1 - [(1 - p)^n]$. This probability increases steadily toward 1 as $n$ increases, so long as $p$ exceeds zero. Thus, in this simple setting, the performance of subunits and units need not be closely connected.

*From Tunnel Vision to Innovation*     Model II's image of organizational change is of sluggish entities dominated by a single way of thinking. Yet modern bureaucracies are often populated by different kinds of professionals with distinctive mind-sets. What is commonplace to one professional culture may be dramatically new—even bizarre—when introduced to an agency dominated by another profession. Observe, for example, the clash between civil engineers and environmental analysts in the Corps of Engineers (Taylor 1984). The questions raised by new environmental analysts about water projects, though alien to the engineers and the corps, were standard in the community of environmental analysts. Or consider the idea of vouchers for schools. To a professional with a background in educational administration, vouchers are a dramatic departure from the status quo. But to an economist specializing in education policy, they are an obvious extension of the basic principle of market competition. In both examples, innovation is based on straightforward applications of professional expertise; heroic mental efforts are not required.[9]

*The Injection of Expertise*     In some domains so much is known, and known explicitly, that it can be codified in books. Sociologists call such domains professions.[10] Becoming an expert means relying less on ordinary folk heuristics, with their attendant biases, and more on scientifically based inferences, with their lower rates of error (Nisbett and Ross 1980; Hogarth 1987). As Nisbett and Ross replied to a colleague who, having read a draft of *Human Inference,* asked, "If we're so dumb, how come we made it to the moon?"

Humans did not "make it to the moon" by trusting the availability and representativeness heuristics or by relying on the vagaries of informal data collection and interpretation. On the contrary, these triumphs were achieved by the use of formal research methodology and normative principles of scientific inference. Furthermore . . . no single person could have solved all the problems involved in such necessarily collective efforts as space exploration. Getting to the moon was a joint project, if not of idiot savants, at least of savants whose individual areas of expertise were extremely limited . . . Finally, those savants included people who believed that redheads are hot-tempered, [and] who bought their last car on the cocktail-party advice of an acquaintance's brother-in-law. (1980, p. 250)

Over the past century, bureaucracies have increasingly relied on other formal institutions—universities and colleges—to provide them with people who enter their organizations already equipped with a great deal of relevant knowledge. This process has been accelerated by organizational innovation, from the creation of think tanks (e.g., Brookings in 1916) to schools of public policy (e.g., Berkeley's Graduate School of Public Policy in 1969). The former directly injects factual analysis into parts of the policy process. The latter produces human capital, embodied in well-trained students, with the same end in mind. Agencies can bring one or both on board.

Some of the above examples pertain to relatively uncontroversial matters.[11] What about politics and expertise? Some readers will think of conflictual situations in which either truth wasn't spoken to power or it was spoken but not heard (Wilensky 1967). This happens all too often. It is not surprising. When someone powerful stands to lose by facing facts, a standard option is to suppress or at least ignore them. Anyone familiar with political history knows that this has been going on for a very long time. What is relatively new is the routine injection of expertise and professional judgment into political institutions. What is surprising is that it happens at all.

How this came about, how some political institutions have gradually modernized in Max Weber's sense, is a topic of great importance. I can barely scratch the surface here, but omitting the topic entirely would leave too big a hole.

### The Long Run: Modernization

Over a century ago Max Weber offered a simple but important hypothesis: modern organizations are more effective than traditional ones. (Today we're more cautious and would add "on average." I'll use that variant here. It's more accurate, given luck's role in life. And though a weaker claim, it's still important.)

The Weberian program should be central to the comparative study of political institutions, but my sense is that its influence has fallen in the past few decades. I'm unsure why. Perhaps it is politically incorrect to say that one organization is more modern than another and is more polite to say that they are merely different. Perhaps postmodernist currents in the social sciences have made us less confident: what is "modern" anyway?

Ignoring this distinction is a huge mistake, for applied as well as basic social science. It amounts to turning one's back on knowledge. To be sure, there are costs in embracing false claims, and political science has had its share of those. But there are also costs in rejecting accurate ones, and it is unlikely that a sensible way to resolve this type 1–type 2 error trade-off is to suppress one mistake completely. And that there is real knowledge in many of our institutions is beyond doubt, except for complete epistemological skeptics. Few people are that.[12]

It is actually rather easy to answer the question, What is a modern organization?[13] Weber (1946), Parsons (1964), Stinchcombe (1965, pp. 145–46) and Udy (1965) have offered the following dimensions. The more these statements hold, the more modern the institution.[14]

(1) *Meritocracy.* Achievement criteria (what people do) are emphasized over ascriptive criteria (who they are).

(2) *Specialization.* Modern organizations are special-purpose entities (Stinchcombe 1965, p.143).

(3) *Scientific culture.* This unpacks into two components: content (warranted knowledge) and processes for producing new knowledge.[15] Since agencies work on practical problems, the content of scientific culture will typically be applied research, not basic research.

The criteria are related both empirically and theoretically. The empirical connections have been examined (Udy 1959, 1962), though less than one would have thought.[16] Theoretically, it is easy to argue as to why, for example, meritocracy and scientific culture are causally connected. The more an organization is driven by ascriptive criteria, the more identity counts and the less what one knows matters. Scientific cultures revolve around knowledge. (It is again important to emphasize that these are scales. Obviously, the German equivalent to the Manhattan Project was partly ascriptive: Jews, for example, were ineligible. But one also had to be a capable physicist or engineer. Similarly, physics or chemistry departments that discriminate against women are only partly modern.)

The criteria are correlated with other properties that we aren't using here to define modernity but that turn out to be characteristic of modern institutions and societies. Stinchcombe was blunt: "Any one of the numerous ways of dividing societies into 'modern' and 'traditional' gives the same result: wealthier societies, more literate societies, more urban societies, societies using more energy per capita, all carry on more of their life in special-purpose organizations, while poor, or illiterate, or rural, or technically backward societies use more functionally diffuse social structures" (1965, pp. 145–46). Although some oil-rich states are an exception to these patterns, they are strong enough to indicate that the distinction between the traditional and the modern is a powerful one: once we know that a society or an organization is modern, in terms of criteria (1)–(3), we know a lot about it, just as the categorization of the largest sea-dwelling animals as whales rather than really big fish tells us a lot about them.

*Modernity and Organizational Performance*    The main causal argument is straightforward (Landau 1973; Wildavsky 1972). The more modern a bureaucracy is, the faster it learns, on average. The faster it learns, the more rapidly it discards ineffective projects and the more it pursues promising ones.[17] Of course, successful innovations are rare.[18] But if one agency's hit rate is 10 percent and another's is 15 percent, the latter will swiftly dominate the former. As growth economists often point out, the cumulative effects of small annual differences are startling. Normalize the performance of agencies A and B to 100 at year 0. Suppose that A generates successful innovations at a slightly faster rate than B (though most innovations fail in both organizations). Due to this difference, A's performance index improves at 3 percent annually; B's, at 2 percent. Initially the organizations remain close: in year 10 agency A is at 134 and B, 122. But by year 20 A has improved to 181 while B is only at 149, and in about seven decades A's performance index will *double* B's. A small but persistent edge in learning speed cumulates, in historical time, to impressive performance differences.

The performance differences between modern organizations and traditional ones are most evident in domains where agencies compete head to head. (More generally, scholars [e.g., Konrad and Pfeffer 1990] have advanced the hypothesis that the easier it is to measure an agency's performance, the faster will achievement criteria displace ascriptive ones.) Military organizations are a paradigmatic case.[19] To see how striking the impact of modernization can be, it is worth taking a short detour and examining a case: the Japanese naval bureaucracy between 1868 and 1905. This is not intended to be a systematic study of the effects of modernization on public agencies. Far from it: the case-selection involves deliberate sample bias. Despite that, I think it is illuminating.

*The Japanese Navy, 1868–1905*    At the time of the Meiji restoration, the Japanese navy was pathetic. An ill-organized collection of small, poorly armed ships, it probably could have been destroyed by a single British ship of the line. However, if we fast-forward thirty-seven years to the Russo-Japanese war of 1905, we see a major transformation. Although the Russian navy was not equal to the British navy, it was fairly modern, especially its ships (Woodward 1965).[20] Yet its Baltic fleet was crushed by the Japanese navy at Tsushima. What happened between 1868 and 1905? Modernization is often measured in centuries. How did the Japanese transform their navy so quickly?

Based on Weber's concept and associated hypotheses (Stinchcombe 1965), one would expect to find the following.

(1)    Modernization at this pace would have been impossible if literacy hadn't been widespread in 1868.

(2)    The Japanese must have created a professional officer corps. Advancement must have been strongly influenced by achievement criteria: for example, being a samurai from an eminent family would ordinarily be insufficient for advancement (although a competent person from such a family would be fine).

(3)    By the late 1800s one had to know a lot in order to be an able naval officer. And even though specialization would cut down the number of things a new officer would have to learn, there were limits to how much this could be exploited, given the desirability of organizational robustness in the face of shocks to the chain of command during battle. (There are stories that a midshipman wound up commanding a British ship in a battle in the Napoleonic wars because all his superior officers were killed or incapacitated. Complete specialization—disjoint skill sets—would make such a substitution impossible.) Although some of the knowledge would have had to be acquired experientially, some must have been codified. Hence, a naval college must have been established or greatly expanded.

(4) Naval technology in the late 1800s had become quite sophisticated. Learning by doing by a single organization (the Japanese navy) would be much too slow: Japanese officials must have tapped into existing knowledge bases.[21] This illustrates, once again, the importance of cumulative learning.

(I am tempted to say "and so on..." but that would be cheating.)

A quick investigation of the empirical literature (principally Evans and Peattie 1997, but see also Jane 1904 and Schenking 2005) reveals the following.[22]

(1) Although Japan in 1868 was much less industrialized than England or Germany, it was already a highly literate society (Sugihara 2004).

(2) A naval war college was established in 1888. Officers admitted to this institution received extensive scientific and technical training.

(3) Selection into the officer corps was initially mostly ascriptive, based on social class, but competitive exams were introduced in 1871 and by 1901, 34 percent of the academy graduates were commoners.[23]

(4) Systematic evidence was increasingly used in decision making. For example, the navy began to use exercises to adjudicate between competing hypotheses about how to go into battle (Evans and Peattie 1997, pp. 36–37).

(5) Existing knowledge bases were tapped by (for example) hiring British instructors, sending students to the U.S. Naval Academy, and buying foreign ships to jump-start the Japanese shipbuilding industry.

As Landau (1972) and others have argued, good theories reduce surprise in their domains: they reduce the information content of empirical messages. By this criterion Weber was certainly on to something.

Although bureaucratic quality varies substantially, even after one controls for degree of modernity, Weber's hypothesis that the average modern military will usually beat the average traditional one seems like a good bet. And if the difference in modernity is big, the contest can be short and brutally one-sided (consider the Battle of Omdurman in 1898). Part of the difference is obvious—hardware (e.g., guns versus spears)—but Weber's theory tells us also to examine more subtle organizational properties: training, skill at using the more advanced technology, and the coordinated execution of routines. Further, advanced military technology is itself the product of modern institutions: specialized firms embedded in market economies.

These performance claims are weaker than hypotheses about progress—that is, naive hypotheses that institutions move only in one direction, from traditional to modern.[24] This is now a straw man: it's hard to think of a contemporary social scientist who makes such claims. Counterexamples to the naive hypothesis of ever-increasing modernization are easily found. For example, the German army in World War II was more ascriptive than its predecessor in the World War I: German Jews were largely excluded by the former but not by the latter. Closer to home, the administration of Bush II was notoriously dismissive of science and systematic evidence, and it would be surprising if this attitude hadn't damaged parts of the permanent bureaucracy.

*An Upper Bound on Bureaucratic Performance: What Robert Merton Might Have Said to Amos Tversky*     What can we expect from modern bureaucracies? How much can we reasonably expect a scientific orientation to ameliorate the cognitive constraints of individual decision makers?

These questions can't be answered with precision, but in his brilliant book *Making Bureaucracies Think* (1984), Serge Taylor hypothesized that we can estimate an upper bound on the amount of improvement: it's the amount of improvement produced by scientific structure and culture in basic research institutions (principally universities). His argument is almost deductive.

(1)   Scientific institutions have produced basic knowledge about the world at an amazing rate over the past few hundred years. No other kind of organization comes close.

(2)   Universities and other basic research organizations enjoy several enduring advantages over governmental agencies and most other institutions that work on practical problems.[25] Two are especially important. First, there is little time pressure: scientists can disagree indefinitely. Second, universities are quite autonomous and have highly specialized goals for generating or transmitting knowledge. Hence, typically there is little interference from other societally important goals, such as religious commitments or political ideologies.

(3)   These advantages cannot be enjoyed by public agencies, given the practical work they do. The press of time, for example, is a common

theme in pragmatically oriented public policy books (e.g., Behn and Vaupel 1982).

(4) Because these advantages are important causal factors—they help to explain point (1)—and because they are stably absent from public agencies, Taylor's conclusion follows.

Having realistic aspirations is vital. As the models in chapters 3 and 5 showed, utopian standards can be destructive: they prevent us from sticking with good or even excellent solutions.

So, the discoveries (!) that scientists care about status and glory (Watson 1968), cling stubbornly—too stubbornly, in some cases—to cherished ideas, and even occasionally publish fraudulent results remind me of the great line from the movie *Casablanca*: "I am shocked, *shocked*, that there is gambling here!"—said by an officer who pockets his gambling-based bribe immediately after uttering those words. None of this is surprising. These are human beings, subject to fundamental psychosocial regularities: few people are saints; most of us care intensely about status.

So it is not at all surprising that Watson and Crick wanted to beat Linus Pauling to the prize of figuring out DNA's structure. What is surprising, and historically unprecedented, is that scientific institutions are arranged to generate knowledge at phenomenal rates despite the fact that the relevant individuals are in many ways quite ordinary. True, these scientists are smarter than average, but they're morally ordinary and they're fallible. Marty Landau put it well in his wonderful paper "Objectivity, Neutrality, and Kuhn's Paradigm" (1972). In the following quotes, the first paragraph states the problem, and the second conjectures how this problem not only might be ameliorated but is in fact routinely ameliorated by scientific institutions that already exist.

> No one, save perhaps a tyro, suggests that scientific habits of mind are incompatible with passionate advocacy, strong faith, intuitive conjecture, and imaginative speculation. All of us, scientists included, are subject to countless influences so well hidden as to be uncoverable either by socio- or psychoanalysis. To transform a scientist into that fully aseptic and thoroughly neutral observer of legend is a virtual impossibility. There is no doubt that "there is more to seeing than meets the eyeball"; that what we see is "theory-laden." We can admit out of hand there is no such process as "immaculate perception." Arguments, therefore, which seek to sustain objectivity by predicating neutrality are doomed to fail. (Landau 1972, p. 44)

There's little doubt about the magnitude of the problem identified by this paragraph. On the contrary, in the past twenty years the

Tversky-Kahneman program of heuristics and biases has provided compelling evidence that humans make mistakes in a wide variety of cognitive tasks. (Indeed, it's remarkable how accurate were Marty's views about cognition: the paper I quote from was written before the heuristics-and-biases program was launched.)

Now turn to Marty's hypothesis about the solution. In brief, his argument is that scientific culture—the procedures, norms, and current body of knowledge—and the structure of scientific institutions—for example, how authority is organized—greatly ameliorate the inevitable cognitive shortcomings of individual scientists.

> The entire system of science is based on a variation of Murphy's Law—the prime assumption that any scientist, no matter how careful he may be, is a risky actor; that he is prone to error; that he is not perfectible; that there are no algorithms which he can apply so perfectly as to expunge any and all biasing effects. Accordingly, all his proposals must be subject to error-correcting procedures. The goals of the enterprise [i.e., the generation of knowledge] demand a network of highly redundant and visible public checks to protect against the inclusion of erroneous items in the corpus of knowledge. Such networks are institutionalized control procedures which continually subject "all scientific statements to the test of independent and impartial criteria": not men [or women], but criteria, for science recognizes "no authority of persons in the realm of cognition." (pp. 44–45)

The views of this paragraph are rarely encountered in the Tversky-Kahneman literature.[26] This is to be expected. Most of the people working in that tradition are psychologists, not social scientists. Psychologists, trained to think at the microlevel, are less inclined to contemplate institutional solutions.

Consider the audaciousness of Marty's central hypothesis: "the entire system of science is based on a variation of Murphy's Law." This is a provocative claim, but it's also a bit of a tease: how would one test it? Let me give a specific, testable version of the hypothesis: For every major cognitive bias uncovered thus far by the Tversky-Kahneman program, we will be able to find scientific procedures or structural properties that ameliorate it. Further, the better the science—the stronger its corpus of knowledge—the more institutionalized are those procedures or structures.

This is a strong claim: "for *every* known major bias there exists an ameliorating scientific procedure." So, of course, it may be wrong. That's fine; as a Popperian might say, any theory that can't be wounded by evidence was never alive to begin with (Platt 1964). And of course no social science theory that's really alive is completely accurate. But my hunch is that there's a lot of truth to this one.

Let me put some flesh on these bones by giving an example of a cognitive problem and a scientific procedure that ameliorates it.[27] The problem is called the confirmation bias: people tend to search for evidence that supports their hypothesis. They do not look for disconfirming evidence. Apparently people can suffer from this bias even when they have no strong motivations about the hypothesis in question; cold cognition suffices. But if the thinking is hot, if we treasure out hypotheses (and our favorite hypotheses are our mental children, so treasure them we do), then we expect the confirmation bias to intensify.

This bias is the problem context: it's prudent to assume that it affects everyone, nonscientists and scientists alike. So the question is how to control this natural tendency. Consider random sampling. One of its valuable features is that it makes us encounter negative data. Opinion researchers may have very strong views about, say, the place of religion in public affairs, and may hypothesize that most of their fellow citizens share their views, but a genuinely random survey allows them to be *unpleasantly surprised.*

This is just one example. If Marty's audacious hypothesis is correct, there are years of work to be done on the organization-theoretic foundations of science and the growth of knowledge. Cognitive psychologists could join forces with sociologists, historians, and philosophers of science on this topic. The former would identify the problematic mental tendencies; the latter would test Marty's hypothesis by seeing whether scientific systems have procedures that combat the problems identified by the psychologists.

## THE BAD NEWS: HOW INSTITUTIONS EXACERBATE BOUNDED RATIONALITY

Because of our cultural fixation on heroic individuals trying to solve problems despite the clumsiness of their bureaucracies, this topic has been much studied and is fairly well understood.[28] Hence, I will be brief.

It should be acknowledged from the start that many problems are due to agency issues, or, in the case of dictators or divine rulers, the fact that the leaders aren't agents of most people at all. (The very notion that a godlike pharaoh, for example, could be regarded as an agent of his subjects was probably literally unthinkable—an interesting illustration of the cognitive roots of certain agency problems.) These incentive issues are tremendously important.

They can also interact with cognitive constraints in significant ways. If speaking truth to power is dangerous—for example, it is known that a country's dictator shoots people who bring him bad news—few people will dare to do so. Then the dictator's beliefs (about an opponent's strength or a harvest's bounty) can be wildly off the mark and can remain uncorrected for long periods of time—until reality comes crashing in.[29] Tough guys create this problem for themselves.[30] They can even cling to their beliefs after such a crash—literally. After a Luftwaffe general told Hermann Goering that an allied fighter plane had been shot down in Germany, much farther from its base than the Reich marshall had believed any fighter could travel, Goering gave the subordinate a direct order: the plane wasn't there (Speer 1970, pp. 289–90)!

As this example illustrates, incentives are not the only issue. Institutions can exacerbate bounded rationality problems in several important ways.[31]

(1)   *Truth muzzled by power.* As noted, many organizations frequently suppress bad news (also known as error-correcting feedback). Organizational leaders can then fool themselves into thinking that they know things that they don't. This self-delusion can have lethal consequences.

(2)   *Response to threats.* Extreme centralization robs an organization of vertical redundancy and adaptive capabilities. For example, following the Nazi invasion of the Soviet Union, Stalin may have suffered something like a nervous breakdown. Because of the hypercentralized nature of the system, Soviet field commanders were paralyzed by Stalin's collapse: afraid to do anything without the dictator's okay, they failed to improvise effectively in the early days of the Nazi onslaught. The combined effect of leadership breakdown and a hypercentralized command structure was an unprecedented disaster: millions of Soviet soldiers were killed, wounded, or captured in the first few months. A more decentralized institution would have dampened the effect of the random shock of the leader's incapacitation; the Soviet system amplified the shock.[32] Dictators reveal in the sharpest way tendencies that most hierarchies exhibit in varying degrees.

(3)   *Response to opportunities.* Extreme centralization can also increase the probability that organizations will fail to seize opportunities. For example, economic historians have been puzzled about why Western economies grew faster than Imperial China's, possibly beginning as early as 1500 (Maddison 2005, p. 18). In the middle ages, China had many features that would have led one to predict that it would develop faster than the West. Why didn't it? One

hypothesis (Landes 1999) is that under the centralized authority structure of Imperial China, a few emperors and a handful of advisors failed to see the opportunities inherent in technological development and trade with other countries.

(4) *Homogeneity and conformity.* Condorcet jury theorems and related aggregation ideas (Surowiecki 2005) rely on the exercise of independent judgment. Some of these procedures, especially those related to the effectively redundant generation of new alternatives, rely on diversity (Bendor 1985; Page 2007a). Ensuring that these properties hold in a formal organization is not a simple matter. Even modern organizations can be plagued by judgment that is both homogeneous and interdependent. Conformity is often to blame. Even without overt incentives, human beings often conform to their social environments, sometimes without any awareness of doing so (Wegner and Bargh 1998). This is, of course, often benign: when one is uncertain about which course of action to follow, imitating others can be quite sensible (Hurley and Chater 2005). Indeed, some scholars (e.g., Richerson and Boyd 2005) have argued that without a strong predisposition to imitate, humans could not have produced the phenomenal cumulative learning that characterizes our species. The danger, then, is that the unthinking use of the imitation heuristic produces correlated similar beliefs when uncorrelated heterogeneous ones would work better.[33] If these processes are taken for granted, their effects can be invisible.

(5) *Organizational complacency.* Elite and once-elite institutions easily grow complacent. This reduces aspiration levels, which in turn reduces the propensity to search for new alternatives (Simon 1956; Cyert and March 1963). This informal organizational atmosphere reduces the probability of innovation and hence retards progress.

(6) *The cult of leadership.* No one is infallible: not the pope, not George Bush or Barack Obama, not Moses, Jesus, Mohammed, or Buddha. No one. This is not a theological claim. It is an empirical one: every human being makes mistakes. More, the cognitive wetware of leaders is basically the same as that of most people. Some are more shrewd than average; some have better insights on the use of power or even the efficacy of particular policies. All this is well within the normal range and completely consistent with the hypothesis that basic cognitive parameters (e.g., the capacity of working memory) have roughly similar values up and down political hierarchies.

These claims are utterly banal when written down, but they run across more than a few cultural grains. Many premodern systems have sought all-wise leaders, saviors or men (they've almost always been male) on white horses who will lead their people unerringly. Modern systems are not immune to similar temptations: we are tempted to look for the CEO with the golden touch or the expert with all the answers (Surowiecki 2005, p. 36; Camerer and Johnson 1991). We continue to overestimate what individuals can do.

We also overestimate what they have done. Real meritocracies are tough: accountability for performance never stops. This is not only to satisfy the impatient 'what have you done for me lately?' question; ongoing account-ability means that institutions will get bigger data sets on people's per-formance; hence they will more accurately estimate their competence. As Surowiecki points out (2005, pp. 219–20), in any contest *someone* will do better than the rest; how much is due to durable skill and how much to luck is hard to sort out unless we observe more performances.[34]

Moreover, real expertise is domain specific (Ericsson and Lehmann 1996). There is no known exception to this pattern. This has significant implications for leadership. First, if some leaders are real experts, with pow-erful problem-solving skills, these must be confined to a specific domain; for example, a person might be a brilliant submarine commander. Second, to the extent that a leader is a generalist, his or her problem-solving skills are weak. (For an analogous hypothesis about problem-solving methods, see Newell 1969, pp. 372–73.)

Whatever we discover about the breadth and depth of leaders' abili-ties, a modern polity holds everyone accountable for his or her on-the-job performance. The less that retention and promotion decisions are based on merit, the less modern the system. (These are analytical claims.) The selection procedure need not be democratic to qualify as modern (meri-tocratic), but certain kinds of succession methods are clearly traditional: strict monarchy, for example. Being the king's eldest son is a prototypically ascriptive property. Being regarded as wise—or worse, infallible—because of this property is thoroughly inconsistent with an institutional perspective that takes bounded rationality and, relatedly, fallibility as cornerstones of political culture (Kassiola 1974). As Judge Learned Hand said, "The spirit of liberty is the spirit that is not too sure that it is right" (1960). (See also Grofman and Feld 1988 and List and Goodin 2001.)

Some may say that the province of experts is means, whereas that of leaders is ends. But the modern view, I believe, is that no one has special authority on value premises.[35] It is hard to improve on the prescription, implicit in Dewey (1927), for generating factual and valuational premises in modern polities: science should provide the former; democracy, the latter.


## CONCLUDING THOUGHTS

About a half century ago, Herbert Simon and Charles Lindblom launched an intellectual revolution in the general study of decision making. Aaron Wildavsky, Marty Landau, and a few other scholars applied many of the key ideas to the study of politics and policy making.

Revolution is a strong word, but I'll defend it. Here's the gist of the argument. (1) It took economists and decision theorists scores of decades to refine the idea of (fully) rational choice. (2) This concept is central both to basic and applied economics and hence to those parts of public policy analysis—and they are numerous—that are essentially applied economics. (2) Economists have devoted well over a hundred years to working out the implications of assuming that decision makers are fully rational. Obviously, the idea has been both incredibly fertile and incredibly broad: I doubt that there are any known limits to its domain of application.[36] (4) The core premise of the bounded rationality research program is that many real-world choices are significantly affected by decision makers' cognitive constraints.

These points together imply the following conclusion: the BR research program does not only critique what Imre Lakatos (1970) called the "hard core" of a tremendously important incumbent research program; it now offers an alternative (and, at the level of theories and models, alternatives). Given the fertile implications that have been teased out of the idea of fully rational choice over the past century, it is virtually guaranteed that the idea of boundedly rational choice will also generate a huge set of implications. Many of these—not all—will conflict with the incumbent's.[37]

This basic science debate will go on for a long time: decades, I expect.[38] Meanwhile, life goes on, both in the applied sciences and in their domains (i.e., real-world problems). Applied scientists shouldn't hold their breath and wait for the debate's end: basic scientists studying decision making should be pushed to offer hypotheses about the practical implications of their theories.

But let me also urge my friends in policy schools to be patient. As Marty Landau (1977) emphasized, the term *applied science* presupposes that there is a body of fundamental knowledge—a basic science—to be applied. And policy analysts, in deciding whether to use a finding from behavioral decision theory, are in the FDA's position: they can be either too patient, and fail to apply an idea that is ready, or too impatient, and apply something prematurely.

Thus, there is a reflexive problem: what are realistic aspiration levels regarding patience? I have no answer to this question: figuring out what are reasonable aspirations in such contexts is very hard. But then, no one said that it would be easy. Fortunately, it's also interesting.

# Notes

PREFACE

1. An eminent physicist is said to have critiqued a paper by saying "This paper is so bad it's not even wrong."

2. Part of the Wildavsky lecture now appears in the concluding chapter.

3. Thanks go to Lee Friedman for urging me to orient the final chapter this way.

4. Hence, I have kept plural pronouns ("we" and so forth) in all the coauthored chapters. Switching to the first person singular would be misleading.

CHAPTER 1

1. This was probably not Tversky and Kahneman's intention. Their idea was to map the subtle mental framing that can cause cognitive illusions, as studies of visual illusion have helped us understand the visual process (Tversky and Kahneman 1986, p. 260). But this original goal was often forgotten by many of their followers, who seemed determined merely to document the "illusions, foibles, flops, and bloopers to which ordinary people were apparently prone" (Gilbert 1998, p. 21). For an attempt to get the work back on the original track, see Griffin and Kahneman (2003).

2. Readers interested in the history of social science will be struck by the fact that both Wildavsky and Simon were impressed by the cognitive and political complexities of governmental budgeting. Indeed, in his autobiography Simon credited his early research on budgeting in Milwaukee (1935) for stimulating

his thinking on the fundamental aspects of bounded rationality (1991, p. 370). The parallel to Wildavsky's thinking, so evident in chapter 2 of *The Politics of the Budgeting Process*, is remarkable. (See chapter 2 for an extensive quote of Simon's reflections on how this study of budgeting influenced his thinking.) Simon's 1935 paper was unpublished, and though *Administrative Behavior* quotes from it a bit (1947, pp. 211–12), the quote does not reveal the new research problem that he'd discovered. So it seems that Wildavsky's response to federal budgeting was independent of Simon's response to Milwaukee's. However, Aaron was probably predisposed to think this way, given his exposure to the problem-solving orientation of Lindblom and Simon's theoretical essays of the 1950s.

3. I suspect that some of the critics of theories of budgetary incrementalism made exactly this error and were bothered by what they saw as the mindlessness of incremental decision processes.

4. It is important to recognize that a problem representation can be evoked automatically; it need not be the result of conscious deliberation. Indeed, in line with current "dual-process" theories of judgment and choice (Kahneman and Frederick 2002), I suspect that aspiration-based behavior is often rapid and semiconscious. If so, this could help us discriminate empirically between optimal search theory, wherein the optimal stopping rule is (under some interpretations of the theory) consciously calculated, and satisficing.

5. I'll restrict my comments to the standard notion of aspiration level, which dichotomizes payoffs into two distinct subsets. There are probabilistic versions— e.g., the probability that an alternative is acceptable is increasing in its payoff—and versions that generate more than two subsets of payoffs, but I ignore these here. The two sets I describe are a typical formulation.

6. The idea of *problem representation* (sometimes called "problem space") is a major concept in the cognitive psychology of problem solving. See, for example, Anderson 1995, pp. 239, 262.

7. In Simon's verbal theory the attribute space could be multidimensional. Most formal models presume a unidimensional space, but there is no in-principle barrier to multidimensionality. Indeed, since the information-processing complexity rises with the dimensionality of the attribute space, it seems more plausible that a decision maker would resort to a simple search heuristic such as satisficing for multidimensional problems than for unidimensional ones.

8. A toy example might clarify the point. Suppose an agent knows that there are five possible options that can be uncovered via search. The options are worth 1, 2, 3, 4, or 5 dollars; every time the agent searches, each of these options has an equal probability of turning up. (These assumptions keep the problem simple and tractable.) The cost of search is 50 cents. Suppose the agent is risk neutral: he or she wants to maximize the expected monetary payoff. What is the optimal stopping rule? Clearly the agent should stop if a 4 dollar option turns up: the expected marginal gain from searching further is only $(0.2) \times (1 \text{ dollar}) = 20$ cents, which is less than the 50 cent cost of searching further. But he or she should not stop if a 3 dollar option turns up, for then the expected marginal gain of searching again is $(0.2) \times (1 \text{ dollar}) + (0.2) \times (2 \text{ dollars}) = 60$ cents, which exceeds the search cost. So the optimal stopping rule in this case is,

"stop as soon as you get an option worth at least 4 dollars." The ensuing behavior of this optimal searcher will be observationally indistinguishable from the behavior of a satisficer with an exogenously fixed aspiration level of 4 dollars.

9. Behaviorism in psychology should not be confused with behavioralism in political science. Behavioralists are perfectly comfortable with mentalistic concepts, such as political attitudes.

10. Because this seems to be an interesting instance of a "Mertonian multiple" (regarding a scientific discovery), it is worth quoting the report of two eminent learning theorists: "In studies of learned performance, a given reward for a response may have either an incremental or decremental effect upon performance depending on what reward the subject expects or on the range of alternative rewards the subject has been receiving in similar contexts. If a person is expecting a one cent payoff, getting ten cents is going to be positively rewarding; if he is expecting a dollar payoff, then ten cents is frustrating and may have the effect of a punishment. Effects such as these have been observed with animals as well as men . . . They can all be interpreted in terms of Helson's concept of adaptation level. The rewards obtained over the past trials in a given context determine, by some averaging process, an internal standard or norm called the adaptation level. *Each new reward is evaluated in relation to this adaptation level, having a positive influence on behavior if it is above the norm, a negative influence if it is below*" (Hilgard and Bower 1966, p. 486; emphasis added).

11. It can be shown (Bendor, Kumar, and Siegel 2004) that this view is, in fact, justified for a wide range of aspiration-adjustment processes.

12. Both in their original presentation of the theory (Kahneman and Tversky 1979) and in many subsequent papers, Tversky and Kahneman clearly stated that "an essential feature of [prospect theory] is that the carriers of value are changes in wealth or welfare, rather than final states" (p. 277). Indeed, in that paper they referred to this claim as "the cornerstone" (p. 273) of their theory.

13. Students often find this hard to believe. For them the notion of a gain, above and beyond a reference point, is both so fundamental and so intuitive that it strains credulity to hear that it is not a building block of utility theory.

14. More precisely, the reference point is often assumed to be the status quo situation; the corresponding aspiration level is the agent's payoff arising from this position.

15. For an excellent introduction to this new field, see Kahneman, Diener, and Schwarz 1999 and Kahneman and Krueger 2006. For policy implications of hedonic psychology, see Layard 2006.

16. Some may feel that this line of reasoning would lead to a dangerous degree of subjectivity in policy evaluation and that it would be better to rely on objective proxies such as dollars gained and spent. I think that this response amounts to goal displacement. If one embraces utilitarianism, as probably most policy analysts and professors of public policy do, then the fundamental object of interest should be human happiness. This variable is *intrinsically* subjective; if human beings are so constructed that they are content when payoffs exceed aspirations and dissatisfied otherwise, then that is a fact about the world that policy evaluation should take into account.

17. I don't know if such a model exists. If not, don't worry: it will.

18. In his famous critiques of the "comprehensive" method of problem solving, Lindblom argued that for most policy problems the real choice is not between incrementalism and the comprehensive method (the latter is infeasible) but between more and less effective noncomprehensive methods. It was a shrewd insight.

19. I confine my remarks here to fields in which expertise clearly exists: chess, structural engineering, cryptography, etc. (An interesting difference between the heuristics-and-biases branch of BR and the problem-solving branch is the stance taken toward expertise: whereas scholars in the latter are interested in explaining how genuine experts manage to perform their information-processing feats, scholars in the former have been uninterested in that question and are more keen on showing that *despite* their expertise specialists still make inferential mistakes, just like the rest of us.)

20. Some deep work in normative decision theory needs to be done to clarify what we mean when we assert that a heuristic is "pretty (reasonably, very, quite, etc.) good." Clearly these assessments are related to having reasonable aspiration levels, but they may turn out to be merely restatements of the problem.

## CHAPTER 2

1. Of course, as social scientists have discovered, identity can be subtle and multifaceted. So it should not surprise us that when asked to choose between being regarded as an economist or as a political scientist, he easily picked the latter (Simon 1999a, p. 112), saying "My tribal allegiance is to political science." But this is much weaker evidence than what he worked on or how he saw his professional identity.

2. The emphasis here is on a *short* list of *essential* properties. Relative to psychology, political science is a complex macrofield. Hence, to keep their theories tractable and reasonably general, political scientists should be ruthless about microassumptions, postulating only the most important information-processing properties of individuals. (This strategy parallels the early approach taken by scholars such as Newell and Simon, who advocated that cognitive scientists should concentrate on the information-processing level and take into account only the most fundamental properties of *their* microfield, neurophysiology.)

3. The connectionists' claim that lower-order processes (e.g., perception) involve a significant amount of parallelism is taken seriously in the literature.

4. For example, in their well-known 1986 paper, Tversky and Kahneman say, "The present results and analysis—particularly the role of transparency and the significance of framing—are consistent with the conception of bounded rationality originally presented by Herbert Simon (see, e.g., Simon 1955, 1978; March 1978; Nelson and Winter 1982)" (pp. 272–73). Further, they recognize the importance of Simon's scissors: "Perhaps the major finding of the present article is that the axioms of rational choice are generally satisfied in transparent situations and often violated in nontransparent ones" (p. 272). Establishing which problem-representations are transparent—that is, cognitively obvious—and which are opaque has been important throughout the T-K line of work.

5. Exactly what *is* "sufficiently" difficult turns out to be more subtle than the Simonian branch had perhaps appreciated. More on this shortly.

6. Lindblom seems to share this perspective, though it is less explicit in his work. See Braybrooke and Lindblom 1963 (pp. 41–57, 66–79) for a discussion of the relation between cognitive capacities and task complexity.

7. Another empirically based aspiration level for cognitively demanding problems comes from comparing the performance of humans to that of computers. Though the latter are artificial, they are real information-processing systems, not a theoretical benchmark like the fully rational decision maker of RC. Hence, it is interesting to note that cognitive scientists who have actually tried to build artificial systems that perceive three-dimensional objects, understand natural language, or use (so-called) common sense are greatly impressed by our facility at these and related tasks (see, e.g., Steven Pinker's paean to human abilities in these three tasks [1997, ch. 1]). Although we take such abilities for granted, it turns out to be hard to construct artificial systems that do these things at all, much less at our level: "The first step we must take in knowing ourselves [is to] appreciat[e] the fantastically complex design behind feats of mental life we take for granted. The reason there are no humanlike robots is not that the very idea of a mechanical mind is misguided. It is that the engineering problems that we humans solve as we see and walk and plan and make it through the day are far more challenging than landing on the moon or sequencing the human genome" (p. 4). Again, a comparison to a real entity provides a realistic aspiration level about information processing. A third approach that easily leads to a positive assessment of human information processing is that taken by developmental psychologists who try to figure out how babies grapple with the "booming, buzzing confusion" of their new worlds (see, e.g., Gopnik, Meltzoff, and Kuhl 1999).

8. This naturally spawns the question, What makes certain problems mentally difficult (Kotovsky, Hayes, and Simon 1985)? The answer for chess is obvious: combinatorial explosion. But as the T-K branch has shown, sometimes the answer may turn on subtle issues of framing and mental representations, and so is much less obvious. (On this topic, see Kotovsky, Hayes, and Simon 1985 and also Simon 1979a, part 7.)

9. This perception is not without some justification. The literature has devoted much space to documenting "a long list of human judgmental biases, deficiencies, and cognitive illusions" (Einhorn and Hogarth 1981, p. 54). "The logical approach [comparing people to optimal baseline models] triggered an explosion of research on inferential error, and the list of illusions, foibles, flops, and bloopers to which ordinary people were apparently prone became rather long . . . soon, if there was a mistake to be made, someone was making it in the presence of a social psychologist" (Gilbert 1998, p. 121). Thus, the impression conveyed by the literature, perhaps not only to the "uncritical eye," is that of decision makers who foul things up. After immersing oneself in this literature, the glass probably does look half empty—or worse.

10. A key axiom of RC theories—so basic that it is often unstated—is description invariance: decision makers should be unaffected by different yet instrumentally equivalent descriptions of a problem. (For a discussion of this postulate, see Shafir and Tversky 2002, p. 84.) The work on framing and problem representation is theoretically important because it shows that humans often violate this key axiom.

11. Gigerenzer also misunderstood the importance of transparency in Tversky and Kahneman's work. They have consistently maintained the following variant of Simon's main principle: when normative axioms are transparent (cognitively obvious), people tend to accept them.

12. For example, Dawes's survey (1998) of behavioral decision theory mentions virtually no studies of the performance of experts—except for work in the Meehl tradition, which shows (convincingly) that experts are outperformed by simple linear models in many choice contexts. Hence, even this exception exemplifies the contrast between the two BR branches. The Simonian tradition studies experts to find out how and why they outperform nonexperts. When the T-K branch studies experts, it is primarily to show either that they make the same errors as everyone else or that they are worse than algorithms.

13. The literature on calibration of probabilities, which has studied the superior performance of some experts (weather forecasters are superbly calibrated), is an exception to this tendency (Lichtenstein et al. 1982; Koehler et al. 2002).

14. For a striking example of this omission, see Slovic 1990 ("Choice") and Holyoak 1990 ("Problem Solving"). Though side by side in the same volume, their citations are almost disjoint sets. The former mentions the Simonian branch infrequently; the latter doesn't cite a single work from the T-K branch! Evidently choice and problem solving have little to do with each other.

15. Moe was careful to add that this was not an in-principle limitation of the research program, but others came to hastier conclusions.

16. For example, the literature on learning, now very extensive in theoretical economics, clearly belongs to the general BR program, even though few of these models build on the specific theory of satisficing.

17. The DDW papers, however, contained more than quantitative models. For example, the 1966 paper sketched a verbal theory of how boundedly rational officials make budgetary decisions. In particular, the authors clearly continued to regard incremental behavior as significant due to its properties as "an aid to calculation"—that is, as a heuristic. Thus, the underlying cognitive theory's thrust remained intact.

18. This was no coincidence. The chair of Crecine's dissertation committee was none other than Herbert Simon, and Padgett's other thesis advisors—Michael Cohen and Robert Axelrod—were, like Crecine, well acquainted with the larger BR tradition that was the theoretical base for *The Politics of the Budgetary Process*.

19. After Pearl Harbor, George Marshall had to decide which integer multiple of the War Department's 1941 appropriation he should request. Though he probably thought of only a few alternatives, it is unlikely that he considered anything in the neighborhood of the pre–December 7 status quo.

20. For a detailed analysis of this misunderstanding, see Bendor and Hammond 1992, pp. 311–13.

21. Here RC and BR models can overlap a lot, because the former can always represent the possibility of error via agents lacking required information. (This is, for example, how RC models of Condorcet's jury theorem are constructed.) However, as Simon frequently pointed out, the *cognitive* assumption of perfect information processing becomes even more strained when one assumes that agents are in a stochastic choice environment. And when one adds in strategic considerations—

essential to game-theoretic models of the jury theorem—the assumption of complete rationality becomes still more dubious.

22. See Ting 2003 for an important advance on the redundancy argument. In his model, agents are cognizant of each other's activities. Because this may induce shirking—I may not work if I know that you are standing by, ready to back me up—a strategic analysis of redundancy shows that the earlier engineering analyses overestimated the value of organizational duplication.

23. Experiments (McKelvey and Ordeshook 1990) indicate that subjects can take a lot of time to converge to the median when they are uncertain about voters' preferences. And real parties rarely converge fully (Stokes 1999, p. 258).

24. This closely parallels Lindhlom's argument in *The Intelligence of Democracy* (1965).

25. Because the researchers examined equilibria produced by optimizing agents, RC analyses of party competition have systematically ignored the cognitive difficulty of the agent's choice problem.

26. When I presented an early version of the Bendor, Mookherjee, and Ray paper on adaptive party competitition at a Stanford seminar, Dave Baron asked, "Why don't you try this apparatus on a hard problem?" When asked for an example of a hard problem, Baron replied, "Turnout."

27. I believe that Simon's theory of procedural rationality implies that if/when experts optimize, typically they do so via recognition processes. Heuristic search is a fallback method, used for hard problems; hence, the theory suggests that this method needs luck to produce optimal results. (As Simon's theory of procedural rationality is not a formal argument, proving this conclusion rigorously requires further work.) Thus, his view implies a nuanced replacement for the too-simple belief that amateurs satisfice but experts optimize. As argued in chapter 1, that naive idea can establish unrealistic expectations. If, for example, we help young foreign policy analysts learn how to generate sensible solutions to difficult problems, we will be doing well. (For a penetrating analysis of some problems associated with that task, see Tetlock 2005.)

28. Note the place of satisficing in this overall view. Far from being the heart of bounded rationality, it is merely one of several general-but-weak heuristics.

29. Even foxes, who consistently outperform hedgehogs in Tetlock's impressive study of foreign-policy specialists, do not "hold a candle to formal statistical models" (2005, p. 118).

30. For a clear example of scholars putting different weights on these criteria, see the exchanges between Green and Shapiro and their critics (Friedman 1996).

31. To be fair to Simon, this debate encompasses two very different senses of simplification. One involves ignoring most of what we know about a microlevel. This he endorsed: he studied cognition but ignored neurophysiology. The second involves idealization: making assumptions that we know to be counterfactual in order to make our theories tractable—simple enough so that we can work out implications via closed-form mathematics. The later Simon increasingly objected to this. (In fact, it seemed that he objected to idealization-type simplifications even when these are not known to be counterfactual but aren't known to be true either.)

32. Although this was a useful research strategy for a long time, it now seems to be breaking down—but for benign reasons. Neuroscience is presently generating so many tempting research opportunities that the Newell-Simon strategy is now regarded as too costly.

## CHAPTER 3

1. The fast-and-frugal school stresses this intellectual lineage (e.g., Gigerenzer and Goldstein 1996; Gigerenzer, Todd, and the ABC Research Group 1999; Gigerenzer and Selten 2001), but the heuristics-and-biases camp also mentions it (Tversky and Kahneman 1986, pp. 272–73; Gilovich and Griffin 2002, p. 2; Griffin and Kahneman 2003, p. 167).

2. Although no one has provided necessary and sufficient criteria that identify a rule as a heuristic, two features are often mentioned. First, since heuristics are often only clues or tips—they help solve problems but aren't full solutions (Polya 1945)—they need not be complete plans of action. Hence, they differ significantly from the rational choice concept of *strategy*, which must specify exactly what a decision maker will do in any possible contingency. (For example, reciprocity is a heuristic. Although it can guide behavior in, say, the repeated two-person prisoner's dilemma, it isn't a complete plan of action because it doesn't tell us what to do in period one, when we first meet someone. In contrast, tit for tat, which *includes* the reciprocity heuristic, is a complete plan of action in the two-person prisoner's dilemma.) Second, and relatedly, a heuristic, unlike an algorithm, carries no guarantee that it will find a solution for the task at hand (Rochowiak 2005).

3. There is evidence of a midcourse correction in both camps. For papers on the heuristics-and-biases side that emphasize that identifying biases is merely instrumental to the fundamental goal of understanding cognitive processes, see Griffin, Gonzalez, and Varey (2001); Gilovich and Griffin (2002); Kahneman (2002); Griffin and Kahneman (2003); and Kahneman and Frederick (2002). Advocates of the fast-and-frugal approach have increasingly recognized that under some circumstances even their favorite heuristics may fail or may be inferior to full-rationality strategies: in particular, Gigerenzer, Czerlinski, and Martignon (1999) analyze when a Bayesian method outperforms their well-known Take the Best heuristic. See also the brief discussion of heuristic failure in Goldstein et al. 2001.

4. Naturally, a specific problem-solving tool (e.g., a heuristic) and a human being's entire cognitive repertoire have very different constraints. In particular, if a person has a large repertoire of tools and can switch between them effectively, deploying each in appropriate problem environments, then the ensemble would be much less constrained than any of its parts (Payne, Bettman, and Johnson 1988; Cosmides and Tooby 1992; Payne, Bettman, and Johnson 1993; Todd and Gigerenzer 2000, p. 740). This *adaptive toolbox* perspective naturally leads to important—and still wide open—questions, such as: (1) How large are our mental repertoires? (2) How effectively do we code (diagnose) problems and match solutions to them? (3) How stable are the problems generated by our task environments? We return to these topics in the conclusion. On the heuristic-selection issue, see the comments by Baguley and Robertson (2000), Cooper (2000), Erdfelder and Brandt (2000), Luce

(2000), Morton (2000), Shanks and Lagnado (2000), Shanteau and Thomas (2000), and Wallin and Gardenfors (2000), and the reply by Todd and Gigerenzer (2000).

5. Scholars in the fast-and-frugal school often cite Simon's scissors analogy. But their use of it is incomplete: they pull Simon into the optimists' camp by emphasizing how our heuristics can match up well with our task environments. But part of the analogy's bite comes from the fact that in difficult environments (Shanteau and Thomas 2000), our cognitive constraints will "show through" (Simon 1996). This was the heart of his critique of economic theories of choice.

6. Consider, for example, this stark statement: "*The capacity of the human mind for formulating and solving complex problems is very small compared with the size of the problems whose solution is required for objectively rational behavior in the real world—or even for a reasonable approximation to such objective rationality*" (Simon 1957, p. 198; emphasis in the original).

7. This is not mainly to help us deal with the emotion-laden core—evaluating human rationality—of the "rationality wars." The rush to answer the question, How smart are we? has been premature; first we need to understand cognition better. As in the Tversky-Kahneman analogy with our visual system, one doesn't really understand a mental process unless one grasps both what it can and what it can't do. In fact, since mental processes aren't directly observable, explaining a person's behavior via an unobserved mental procedure is a hypothesis, and for corroborating a hypothesis, *all* of the hypothesized procedure's observable effects—including both positive and negative performances—are relevant.

8. Demonstrating this would be a worthwhile exercise in itself, especially in light of some statements by key protagonists to the contrary (see Gigerenzer 2004, p. 395). But to do that well would take another paper.

9. The writings of the leaders of the fast-and-frugal approach (Todd and Gigerenzer 2000; Gigerenzer and Goldstein 1996; Gigerenzer 2004) show that the concept of satisficing is an important part of their framework.

10. Even Simon's second performance claim, which appears overtly normative, is also linked to the descriptive theory via the hypothesis that if satisficing performed very poorly, one would expect most people to discard it. (This is weaker than the bold argument sometimes used by economists that if a strategy is suboptimal, then people will drop it, and also weaker than the equally bold premise of some evolutionary psychologists that selective forces have produced optimal mental processes. Simon's argument is merely that *transparently* bad alternatives are ephemeral.)

11. Griffin, Gonzalez, and Varey say that "Gigerenzer has developed his own set of judgmental heuristics based on computer simulations. These 'optimal heuristics' (our term) are based on a satisficing model of judgment that goes back to Simon" (2001, p. 218).

12. Readers with optimistic beliefs about fast-and-frugal heuristics should note that Simon had modest expectations about satisficing's problem-solving power: "If the task domain has little structure or the structure is unknown to us, we apply so-called 'weak methods,' which experience has shown to be useful in many domains, but which may still require us to search a good deal. One weak method is *satisficing*—using experience to construct an expectation of how good a solution we might reasonably achieve, and halting search as soon as a solution is reached

that meets the expectation" (1990, p. 9, original emphasis). See also Newell 1969, pp. 377–80.

13. "An organism that satisfices has no need...of complete and consistent preference orderings of all possible alternatives of action" (Simon 1957, p. 205; see also Simon 1955, 1996, p. 29).

14. This property continues to hold if the decision maker's feedback about payoffs can be incorrect. Indeed, assuming that feedback is accurate makes it *harder* to establish that the optimal action is unstable. In general, incorrect feedback naturally makes it more difficult for a decision maker to stick with the optimal alternative.

15. For a discussion of how unfriendly environments can hurt the performance of fast and frugal heuristics, see Shanteau and Thomas 2000.

16. This problem does not vanish if aspirations are endogenized. I won't give the details here (see Bendor, Kumar, and Siegel 2004), but the following points should help. (1) By the definition stipulated in this chapter, standard bandit problems are ill matched to satisficing. (2) If aspirations are endogenized via weak (general) axioms of adjustment, then we show that aspirations will eventually be sucked into the $(0, h)$ interval, where $h$ is the positive payoff of the two-armed bandit. Then the analysis proceeds as before.

17. Standard bandit problems presume unidimensional payoffs. We can allow for multiple dimensions (so long as each one takes on only two values) but suppress the corresponding notation. This success/failure coding does not imply that a problem is well matched to satisficing. The two-armed bandit has such payoffs yet it is ill matched.

18. Under (A2) agents do not refine their inclinations to try an alternative, unlike, for example, in reinforcement learning models. Though unrealistic, this constitutes a scientifically conservative assumption in attempting to confirm Simon's conjecture: since search does not improve with time, the burden of making progress falls entirely on satisficing. Assuming stationary search makes it harder to show that satisficing behaves sensibly when faced with hard problems.

19. Numerical simulations suggest that property (i) continues to hold if there are more than two actions and search is blind, i.e., new options are selected with equal probability.

20. Moreover, although our assumption about the agent's initial propensities $(p_{1,0} = p_{2,0})$ is very specialized, the results are robust. For example, as the inequality in (i) is strict at every date and the $p_t$'s are continuous in the initial probabilities, this result is not knife-edged: the conclusion holds as long as $p_{1,0}$ and $p_{2,0}$ are sufficiently close.

21. This example reveals a counterintuitive reason for why the neutral start is crucial for part (i) of proposition 1. The intuitive reason for the importance of this assumption is that an initial bias toward the inferior action might persist for a while (if, for example, $\theta$ were close to zero). But the example in the text shows that an initial bias toward the *superior* alternative can (in some parametric environments) also preclude monotonic improvement. This is more surprising.

22. One shouldn't take this analogy literally. An underdamped spring exhibits oscillatory *behavior*: the oscillation described in the text pertains to *probabilities* (or relative frequencies, for a population of agents), not to an agent's behavior. Even in situations when the chance of selecting the better action does increase

monotonically (essentially because $\theta$ is sufficiently small; see Bendor, Kumar, and Siegel 2004), an agent can switch between the two actions; behavior on this sample path need not look smooth.

23. Per claim (5), there may be discarded heuristics that are consistent, but these would be consistently *bad*.

24. Interestingly, Todd and Gigerenzer emphatically reject this optimality claim (2000, p. 771). This marks a significant difference between these fast-and-frugal scholars and some extremely optimistic evolutionary psychologists.

CHAPTER 4

1. This was not the case in the application of incrementalism to budgetary research, where the controversy quickly became quantitative. However, with few exceptions (Padgett 1981), the arguments underlying the quantitative research remained nondeductive.

2. Alternatively, one could assume that the agent finds it too complicated to think about this feasible set in any useful way.

3. Of course, the probability of *generating* an excellent new option may be smaller than that of generating a mediocre one, given a poor starting point. This is a separate matter, to be discussed shortly.

4. Kahneman, Knetch, and Thaler discovered that potential sellers (subjects who first received an endowment of coffee mugs) valued the mugs much higher than did potential buyers, who first received cash: the reservation price of the median seller was about double that of the median buyer (1990, p. 1332). Thus, there were many unconsummated exchanges due to the "endowment effect"—the high valuation of the status quo endowment.

5. If $Q_t$ is symmetrically distributed and the decision maker is exactly as competent as a fair coin, then $E[V_{t+1}|V_{t=1}] = i$: the process's expected value is exactly its current value. Such processes are known as *martingales*. March and Olsen (1984) suggested that policy processes are better understood as martingales than as independent trial processes. Their suggestion is based on the claim that policy making is a path-dependent process (as are martingales), whereas independent trial processes are necessarily path independent. Path dependence does not, however, imply the martingale's very special property that on average one stays where one is. Theorem 2 describes path-dependent processes that are generalizations of martingales.

6. I shall not explicitly model search behavior here but only investigate how different, exogenously fixed search strategies affect outcomes.

7. More precisely, call the distribution of $Q'_t$ riskier than $Q_t$ if, after normalizing them so that they have the same status quo $x_0$, $p(Q'_t > x) \geq p(Q_t > x)$ for all $x > x_0$ and $p(Q'_t < x) \geq p(Q_t < x)$ for all $x < x_0$, and these inequalities are strict for at least one $x$ bigger than $x_0$ and at least one x smaller than $x_0$.

8. Theorem 3 may surprise incrementalists but may not surprise students of optimal search theory (see, e.g., Kohn and Shavell 1974). However, because the latter's work came after Lindblom's, it was unavailable to him. Moreover, in optimal search theory, decision makers are perfectly rational, so even if the work had been available to Lindblom, he might have thought it irrelevant. Indeed, to my knowledge, theorem 3 is the first result to analyze how increasing risk in the set

of alternatives affects a boundedly rational agent and to identify the minimal level of rationality necessary for increasing risk to be beneficial. Because weak competence is a very low standard of rationality, it is interesting that this is all that is required.

9. It is assumed that the status quo is evaluated without error. Theorem 4 does not depend on this assumption, but it simplifies the proof.

10. Since the agent picks the new option if he or she evaluates it as being better than the status quo, $p_{i,i+k;t} = p(i < i + k + \theta_t) = p(0 < k + \theta_t) = p_{j,j+k;t}$.

11. Note that $\theta_t$, and $\theta'_t$ need not belong to the same class of random variables. For example, $\theta_t$ could be uniform, whereas $\theta'_t$ could be normal.

12. The Manhattan Project is an outstanding example of this point: the great advances in theoretical physics made the radical innovation of the atom bomb possible.

13. Because theorem 5 describes the change in the relative benefits of radical search, weak competence is not required: $E[V'_t] - E[V_t]$ increases as (say) $f$ rises, even if $f < b$. Hence, this result holds under more general conditions than does theorem 3, for in the latter, weak competence is absolutely necessary.

14. Note that evaluation of policies based on status quo search is flawless, since $d = 0$. Given this, the probability of correctly choosing a superior new alternative worth $x$ over a status quo worth $x_0$ is just the probability that a normal random variable of mean $x$ and variance $\sigma^2$ takes on a value in $(x_0, -x, \infty)$. A similar property holds for the probability of correctly choosing the status quo over an inferior new option. Because the normal is symmetrically distributed, the probability of choosing correctly always exceeds one-half. Thus, in contrast to the more general error structure of theorem 4, which does not imply even weak competence, this specialized error structure implies strong competence, and the normal's symmetry implies insensitivity to the status quo as well.

15. In this regard Lindblom is part of an old intellectual tradition, going back to Condorcet, that has emphasized that more heads are better than one. For a valuable synthesis of the Condorcet tradition, see Grofman, Owen, and Feld 1983; for some extensions, see Ladha 1992 and Miller 1986. The more general idea of building reliable systems from unreliable parts was pioneered by von Neumann and is now a major field in engineering, much more prominent there than Condorcet models are in political science. Engineering reliability theory was introduced into political science by Landau (1969); see also Bendor 1985 and Heimann 1993. Ting (2003) took the important step of modeling strategic interactions among redundant agents. His results show that engineering redundancy theory often overstates the benefits of duplication in organizations (e.g., because possibly redundant agents might free-ride on each other's efforts).

16. Given the assumptions of equal competence and independent judgments, one can easily prove that under majority rule, a group of weakly competent individuals is itself weakly competent. A similar mapping holds for strong competence. Further, crude, homogeneous, and heterogeneous individual judgments translate into crude, homogeneous, and heterogeneous collective judgments. Hence theorems 1–5 and propositions 1 and 2 remain valid when there are $n$ equally competent and independent decision makers.

17. This comparison is legitimate because figure 5 is based on propositions 1 and 2, and the assumptions of both of these propositions are satisfied by the conditions of theorem 8.

18. "Analysis and evaluation are 'disjointed' in the sense that various aspects of public policy and even various aspects of any one problem . . . are analyzed at various points, with no apparent co-ordination and without the articulation of parts that ideally characterizes subdivision of topic in synoptic problem solving" (Braybrooke and Lindblom 1963, pp. 105–6).

19. All subsequent numerical examples will presume a normalized status quo of $(0, \ldots, 0)$.

20. Another class of problems involves more conflict. For example, suppose the new, equally likely options are $(1, -1, -1)$ and $(-1, 1, 1)$. (Note that dimension 1 is majority inconsistent.) If the decision makers are competent, over time policy domain 1 will get worse and the other two will improve. Here, greater sophistication cannot help: specialist 1 is locked in a zero-sum game with the other two, thus precluding logrolls.

21. Majority consistency does not imply that the policy dimensions are independent. For example, if the following new alternatives are equally likely—$(1, -1, -1)$, $(-1, 1, -1)$, $(-1, -1, 1)$ and $(1, 1, -1)$, $(1, -1, 1)$, $(-1, 1, 1)$—then they are majority consistent. But the dimensions are not independent: the marginal probability of improvement on each dimension is $\frac{1}{2}$, yet the chance that all dimensions improve simultaneously is zero instead of $\frac{1}{8}$. Indeed, the dimensions are negatively related, indicating conflict across the policy domains. Interdependence across policy domains is an important attribute that adds realism to the model. Without it, we would in effect be returned to a unidimensional policy space; the realism added via multidimensionality would be wiped out by an assumption of independence.

Further, independence would mean that the agents would not be in conflict: the chance of generating the best option on one dimension would not be hindered by what was happening on the others. Trade-off frontiers require interdependent dimensions. (For example, with three dimensions, independence implies that the distribution of new alternatives looks like a box in three-dimensional space. The upper northeast vertex represents an alternative that Pareto dominates all others; hence no trade-off frontier exists.)

22. Numerical examples on unbalanced growth indicate that under some plausible conditions the policy domains that do not grow will be hurt by the increased size of the others. As in introducing people specializing in previously neglected policy dimensions (e.g., the environmental effects of building dams [Taylor 1984]), this may be a price that leaders are willing to pay.

23. Note that as the organization increases in size, the number of policy dimensions (hence the number of departments) remains fixed. Hence, under the standard Condorcet conditions, it is the departments that move toward infallibility as they grow larger.

24. Even asserting, as a theoretical premise, that more incremental search generates better average alternatives does *not* establish that incremental search is superior to nonincremental. The reason is that radical search's greater dispersion remains

an advantage. Thus, incrementalism would trade off a better average for smaller dispersion; the net effect is uncertain.

25. This presumes that something like theorem 2 holds, so that the average status quo in the incremental process improves over time. Otherwise, the less often that policy changes, the better.

## CHAPTER 5

1. Some intellectual history may help to explain why these descriptive and normative approaches merged. The study of organizational reliability was first inspired by a metaphorical transfer from reliability engineering (Landau 1969). The latter field naturally focuses on systems that humans have designed and whose properties—especially the probability and cost of failure—we understand well. In such contexts it may be possible to build in optimal amounts and kinds of redundancy. So it is common to see optimization analyses in texts on reliability engineering. Students of organizational design who read such texts are then prompted to carry the analysis of optimal redundancy into organization theory (e.g., Heimann 1997, p. 83). Substantively, however, it is more difficult to design optimally redundant organizations than to design optimally redundant braking systems: we have better data about the latter, and we can more easily test alternative designs. Hence, this part of the metaphorical transfer may be on somewhat shaky grounds.

2. Chapter 3 also explores this tension between exploration and persistence and the ensuing possibility of excessive adaptation.

3. Both models, however, are explicitly dynamic: they depict an organization altering its reliability properties over time. In this respect both extend classical redundancy theory (Landau 1969), which—at least in its mathematical form—was static in nature.

4. See Heimann 1997, chapter 4, for a detailed analysis of the $k$-out-of-$n$ system. One significant difference between his setup and the current one is that the former presumes a specific temporal structure—his series-parallel structure fixes when different agents make their decisions—whereas in the present model these temporal properties need not be specified.

5. A common specification of $L(\cdot)$ is that it equals $\Pr(\text{type I error}) \cdot (\text{cost of type I error}) + \Pr(\text{type II error}) \cdot (\text{cost of type II error})$. The reader might want to keep this functional form in mind as an example.

6. Some texts on adaptive algorithms virtually *identify* the field in terms of how decisions are "made with very little knowledge concerning the nature of the environment" (Narendra and Thathachar 1989, p. 25).

7. Note, however, that theorem 1 does not guarantee that the agency will avoid becoming permanently trapped in a *set* of suboptimal standards. This possibility cannot be ruled out by (A1) because that assumption does not say anything about which new standards the agency will turn to, following dissatisfaction with the current standard. It simply says that *some* search can occur.

8. For this reason, in game theory games for which no player is exactly indifferent between any pair of distinct outcomes are called *generic* games. The idea is that indifference would be broken by any perturbation (no matter how slight) of the original game.

9. Given more specific assumptions about, for example, the median voter's preferences about the two types of errors and about the decision-making technology, one could also easily show that in general the optimal rate of both types of errors exceeds zero. Thus, a fully rational FDA would, from time to time, approve bad drugs.

10. Although it has long been implicit in the verbal theory of satisficing that aspirations are always below the optimal level (per Simon's remarks about searching a haystack for a "sharp enough" needle), in formal models of satisficing this is an auxiliary assumption. It is not part of the axiomatic core of aspiration-based models of adaptation (Bendor, Diermeier, and Ting 2003a, p. 264).

11. A recent FDA head, Mark McClellan, referred to this behavior as "pendulum swing" (2007, p. 1700). As he contrasted pendulum swings to "systematic improvement," the implication was that the former was suboptimal. (Interestingly, McClellan has a Ph.D. in economics, which may lead him to think in terms of ex ante minimization of expected losses rather than sequential adaptation to the most recent failures.)

12. There is a straightforward generalization to multiple actions.

13. Some stationary rules do satisfy our definition of potential optimality, such as the simple rule of randomly choosing a standard in period 1 and sticking with it forever. Such a blindly persistent rule, however, typically performs very badly in terms of a criterion such as minimizing discounted expected losses. Our criterion, which is merely necessary for optimal behavior, is biased toward persistence rather than exploration. Blindly persistent rules sacrifice exploration for persistence; (A1)-type rules do the opposite. (The Gittins index rule resolves this trade-off optimally.)

14. Of course, we do not assert that a real FDA could be so stoic in the face of congressional hearings! This is a normative scheme, not a descriptive one.

15. In recent years many papers, starting with the seminal work of Austen-Smith and Banks (1996), have criticized the classical Condorcet model for assuming that decision makers on a committee or jury vote sincerely and, therefore, independently of each others' behavior and of the voting rule. The assumption in this chapter is weaker: the behavior of decision makers *could* be affected by anticipation of each others' actions or by the voting rule. We assume only that the *net* effect of raising the voting threshold is in the intuitive direction: that it makes passage more difficult. For evidence of this, see the experimental study of Guarnaschelli, McKelvey, and Palfrey (2000).

16. By the same token, of course, such a system is highly *un*redundant regarding type II errors: the agency will mistakenly reject a good proposal if even one decision maker votes to block it. For a detailed discussion of how $k$-out-of-$n$ systems trade off type I and type II errors, see Heimann 1997, pp. 73–86.

17. To gain further insight into (A6), it is useful to consider an example of an adjustment that does not satisfy the property. Suppose the agency's adaptation is "bang-bang": if it discovers a type I error, then it tightens all the way up, to $s_n$; if it discovers a type II error, then it relaxes all the way down, to $s_1$. Because type I errors are maximally likely when the standard is $s_1$, the process is more likely to transit to $s_n$ from $s_1$ than it is from any other standard. Clearly, then, (A6) cannot hold: for example, row 2 of the transition matrix $P$ cannot weakly stochastically dominate row 1.

18. If it has very coarse standards (e.g., either lax or stringent) then it may not be optimal to become more stringent. But at the very least, it cannot be optimal to *weaken* its standards in the face of an exogenous increase in the relative frequency of bad projects. Thus, the completely accurate statement is that the optimal standard is weakly increasing in the relative frequency of bad alternatives.

19. This does not mean that established firms want the laxest possible standards or none at all. Letting quacks push unsafe or ineffective drugs would lower the aggregate demand for the industry. Rather, the claim is merely that the ideal standard of most pharmaceutical firms is lower than that of the FDA (Carpenter and Ting 2007).

20. While the response described by proposition 1 is directionally consistent with optimal behavior, the normative implications of proposition 2 are less clear. *If* the political environment has begun to push too hard, in some normative sense, for project approval, then the agency's leaning against the wind as reported by proposition 2 may be directionally consistent with the maximizing of some welfare criterion. In such circumstances, the bureau would be standing up to overly powerful interest groups that are pressuring the government to pass new projects. But the proposition is silent about whether this "if" holds.

21. Alternatively, and more subtly, the criteria for identifying good drugs could be weakened (Epstein 1997).

22. As was true of the previous result, proposition 3's relation to optimization is not transparent. The fact that the agency has become more informed about the quality of rejected proposals (hence, more aware of type II errors) does not necessarily imply its response to become less stringent is optimal. Proposition 3 should be taken as a purely descriptive claim.

23. Of course, proposition 3 captures only part of this experience. The FDA apparently relaxed only those standards relating to drugs that fight AIDs; others were unchanged. The present model does not represent different standards specialized for different classes of problems. One might, however, be able to incorporate this into an extended version of the model.

24. Note that since Y's error rate is a constant fraction of X's, whatever standard is optimal for X is also best for Y. Thus, the comparison of proposition 4 is clean: the benchmark of optimality stays put.

25. Of course, resolving this tension in a satisfactory manner may often be easy, but this is a very different matter. We are taking the word *optimally* seriously here.

26. Robbins and Monro's paper was prescriptive; they did not claim that their time-dependent learning scheme described real decision makers.

27. We need not become embroiled in the controversies about the political control of bureaucracy here. The problem may be nontrivial for many reasons. The agency may be controlled by the legislature, and for the latter the median legislator is pivotal. Or the agency may be controlled by the president, and his preferences about the ideal trade-off between type I and type II errors predominate. Or the agency may have some discretion, and professional or organizational preferences about the ideal trade-off hold sway. Or any combination of these (e.g., the agency's operational goals are a weighted average of the above) may be true. As long as the combination of external and internal pressures produces, at the end of the day, some standards that are optimal and some that aren't, we're in business.

28. This premise has empirical content: it could be wrong. The agency is an institution, with multiple superiors, so there is no guarantee that it has well-defined preferences. Some—by no means all—rational choice theories of bureaucracy black-box the processes of preference aggregation, treating the agency as a unitary actor with well-defined goals. This subset of rational choice theories of bureaucracy is our focus here.

29. Specialists concerned with the project's reliability interpret this as a loss of vigilance (Perrow, e.g., shares this view), but cutting costs is almost always a legitimate goal.

30. We thank Tom Hammond for some pointed suggestions here.

## CHAPTER 6

1. Research programs are symbolic entities, so it is not surprising that they have fuzzy edges. Consequently, deciding whether a particular theory belongs to a certain program can be difficult. A case in point is the assignment of GCT to BR. This categorization is somewhat controversial, but the originators of GCT clearly believed that it belongs to the BR tradition (Cohen, March, and Olsen 1972, pp. 1–2), and more than a few other scholars seem to agree with them. Bendor, Moe, and Shotts registered some doubts about the accuracy of this categorization (2001, p. 174). I now tend to agree with Cohen and his colleagues—and this chapter's postscript shows how garbage can processes can be modeled quite naturally via bounded rationality microcomponents—but in order to preserve the original argument of Bendor, Moe, and Shotts, chapter 6 retains the doubts expressed in their article.

2. For a reply to the critique, see Olsen 2001.

3. The 1972 paper had garnered 1,847 citations in Google Scholar as of March 31, 2008.

4. This has been reported as an empirical finding by Stephen Weiner (1976), who contributed a case study to March and Olsen's (1976a) edited volume on GCT: "The San Francisco study stimulates a view of the flows of problems, solutions, and energy, when the choice is subject to a deadline, that varies from the theoretical predictions. Problems and solutions, on the one hand, and participants, on the other hand, are intertwined. *Problems and solutions are carried by participants*" (p. 245, emphasis added). But his observation was virtually ignored in the book's theoretical chapters. (For an extended examination of the idea of independent streams, see Heimer and Stinchcombe 1998.)

5. See Padgett 1980b and Carley 1986 for criticisms of GCT's treatment of organizational structure.

6. "Organized anarchies interfere with fundamental tactical principles that depend on tight control and coordination ... adaptations employed by naval commanders include the use of SOPs [standard operating procedures], operational plans, [and] *strong centralized authority acting through a chain of command*" (Weissinger-Baylon 1986, p. 51, emphasis added).

7. A subsequent article (March and Romelaer 1976) does recognize an important role for authority, particularly in allowing superiors to delegate and thereby participate in many decision arenas simultaneously, but this interesting line of analysis has not been integrated into the GCT's larger structure.

8. See, for example, the preceding quote, where the three properties are linked by "and."

9. For a more detailed analysis of the simulation, see Bendor, Moe, and Shotts 1996.

10. Figure 8 is a graphical representation of text information in Cohen, March, and Olsen 1972.

11. Moreover, as we have seen in the analysis of choice behavior in the simulation, to the extent that individuals have goals, they all "want" the same thing: to go to arenas that are close to finishing up. Hence, their interests do not conflict in the slightest.

12. For example, when we say "at the end of period five all five choices are active," this is exactly what the corresponding figure (figure 11) portrays. In the full-scale simulation, all the choices (all ten) are not activated until period ten.

13. The order in which choices and problems enter is irrelevant. Our figures use the following orders: for choices, 3, 1, 5, 4, 2; for problems, 3, 6, 7, 3, 2, 9, 4, 1, 5, 10.

14. In period 2, the decision makers separate from the problems due to a slight difference in how energy balances are calculated. The details of this assumption are unimportant.

15. The old and new problems go to different cans because the computer program assumes that new problems, when they first appear, calculate energy deficits somewhat differently than existing problems do. The details of this assumption are not important.

16. It is possible for agents to separate in simulation variants in which they have unequal amounts of energy. In simulations with unsegmented decision structures, however, all decision makers travel in a single pack more than 99% of the time. Not only do the participants move simultaneously, but the problems do so as well: being synchronized with the movement of decision makers, they all move at the same time. Although the assumption of synchronous adjustment is often used in computer simulation, this is in general a risky modeling strategy. Synchronicity can produce artifactual results that disappear if even small asynchronies are introduced (Huberman and Glance 1993). Furthermore, in the special context of organized anarchies, it is substantively inappropriate to assume that decision makers and problems adjust synchronously.

17. These predictions are reproduced in toto in Cohen and March 1986 (p. 88, table 28). By this point, however, the extended model's auxiliary assumptions not only are unjustified but also are *invisible:* the book omitted the 1972 article's original figures that had represented the auxiliary assumptions.

18. See March and Olsen 1984, p. 746; 1986, pp. 17–18; 1989, pp. 11–14; Mezias and Scarselletta 1994, pp. 658–62; Moch and Pondy 1977, pp. 354–55; and Weissinger-Baylon 1986, pp. 37–40. For exceptions, see Crecine 1986, p. 86; March 1994, p. 201; and Masuch and LaPotin 1989, p. 40.

19. Interestingly, each of these works abandons key components of the GCT. Padgett (1980) and Carley (1986) drop the assumption that participation is fluid. Anderson and Fischer (1986, p. 145) modify the assumption of independent streams: "Individuals are the vehicles which carry solutions and parochial problems to choice opportunities." Most important, all three emphasize the instrumental nature of

individual and organizational action (Padgett 1980b, pp. 585, 591; Anderson and Fischer 1986, pp. 153–54; Carley 1986, pp. 165, 177).

20. We are not sure why this happened. One colleague (Carley, personal communication, 1996) has suggested that the problem is that simulation is not part of the standard repertoire of doctoral students. (See Masuch and LaPotin 1989, p. 38, for other explanations.) Thus, mainstream organization theorists have overlooked technical challenges and new simulations, attending instead to Cohen, March, and Olsen's verbal accounts of their simulation.

21. See, for example, Cohen and March 1986, pp. 193–203; March 1978, 1988, 1994, ch. 5; March and Olsen 1975; and March and Weissinger-Baylon 1986, chs. 1–2.

22. See, for example, March and Olsen 1976a, 1984, pp. 738–41, 743–44; 1989, pp. 11–14; Cohen and March 1986, pp. xv, 91, 195–215; and March 1994, pp. 177–80.

23. This newer work has a postmodernist flavor. The emphases on language and symbols, meaning and its social construction, and the chaos that lurks beneath our delicately maintained interpretations, are all postmodernist themes. These ideas are only touched on in the 1972 article but are given much more attention in *Ambiguity and Choice in Organizations*. Indeed, the latter may well be the first sustained postmodernist study of organizations—a fact recognized early on by Perrow (1986, pp. 137–38).

24. See especially section 4, "Reorganization as Garbage Cans" (1983, pp. 285–87), and the concluding observations (p. 292).

25. For example, the discussion in the 1984 article on "normatively appropriate behavior" (p. 744) is completely consistent with March and Olsen's earlier analysis of how attention is organized (1976b, p. 49).

26. This idea is also central to March and Olsen 1996. See also March and Olsen 1995 (ch. 2) and 2005.

27. The first definition of "rule" in *Webster's Collegiate Dictionary* (9th ed.) is simply "a prescribed guide for conduct or action."

28. To be sure, some institutions (e.g., religious ones) have rules that encode sacred rituals. These differ in many ways from the problem-solving operators analyzed by Anderson. In particular, sacred rituals need not have a "logic of consequentiality," so their logic of appropriateness may be independent of secular notions of instrumentality. But the argument in *Rediscovering Institutions* (1989, p. 160) that "politics is organized by a logic of appropriateness" is unqualified by any distinction between the sacred and the profane. The claim is unquestionably intended to cover ordinary behavior in ordinary governmental institutions. In this domain, the fatal ambiguity identified above—that rule-governed behavior may be infused by the consequentialism of problem-solving operators—is quite evident.

29. In one section (March and Olsen 1989, pp. 30–35), the authors did construct an argument with individualistic foundations. But this was an exception to the book's central tendency, and they made no attempt to introduce or summarize their overall approach via an explicit microtheory.

30. A major exception is Kingdon (1984), whose work is distinguished by a careful empiricism tied to theoretical concerns.

31. For an exception, see Cohen, March, and Olsen 1976, p. 36.

32. For example, in his well-known text on organization theory, Karl Weick applies the idea of organized anarchies to universities: "If that's partially what a university organization is like, then a thick description of that organization will be confusing when it starts to comment about goals ..., technology ..., or participants.... The irony is that *this confusion in the observer's report testifies to its authenticity and not to its sloppiness. Confusion as an indicator of validity is a critical nuance*" (1979, p. 11, emphasis added).

33. "The central idea of the garbage can models is the substitution of a temporal order for a consequential order" (March and Olsen 1986, p. 17). See also March and Olsen 1989, p. 12; Cyert and March 1992, p. 235; and March 1994, pp. 198–200.

34. Evaluation error can easily be introduced: the superior could sometimes select an option that is objectively inferior to some other alternative, as in the models of muddling through (chapter 4).

35. It is methodologically limited as well. If the superior's aspiration is sufficiently low, then chance rules: any alternative can be selected. Thus, the model has no empirical content: it says anything can happen. Hence, because any decision by the boss is consistent with the prediction, any ethnographic description of what happened in this organization is guaranteed to "corroborate" the model. This is a variant of the folk theorem problem identified in chapter 1 by theorem 1. Empirical vacuity can afflict formal models (e.g., the above fixed-aspiration model of satisficing) as well as verbal theories (e.g., the garbage can).

36. Ken Shotts coauthored the postscript, and I thank him for his contribution.

37. This is an exogenous stream of solutions, so here we are guilty of something we criticized in our original examination of GCT. However, in a more complex model one could endogenize the solution stream: lower-level staff, with their own preferences, would generate alternatives for the governing committee.

38. We thank Chris Stanton for doing the simulations.

39. The process is finite because it is represented by a computational model and computers are finite state machines.

40. Visual inspection is corroborated by various quantitative measures. (Summary statistics of a large number of runs provide good estimates of the properties of a limiting distribution.) For example, we computed the average density for a series of concentric circles, all centered at (0, 0). Average density falls as the circles get bigger. Thus, most of the probability mass is indeed centrally located.

41. A good case can be made that this represents maximally fluid participation: each decision maker is just as likely to be absent as to be present.

42. Indeed, we suspect that the distribution of status quo points in period 1,000 will spread out smoothly and monotonically as participation becomes more fluid (i.e., as the probability falls from 1.0 to 0.5): a comparative static result on participation.

## CHAPTER 7

1. The concerns of these groups differ somewhat, due to the difference between basic and applied research. In basic science we want explanations—answers to questions such as why an event happened—and for such questions "as if" formulations (Friedman 1953) won't cut it. For applied work an instrumental approach—don't

question my assumptions, just check my predictions—is fine, since what really matters is whether a given policy has the desired effects. (See Psillos 2000 for a discussion of one of the most important alternatives to instrumentalism—realism—and see Boland 1979 for a defense of Friedman's instrumentalism.) But there still is an obligation to generate reasonably adequate predictions for applied work. And what we consider reasonably adequate will depend on whether there are competing models, for their existence can raise our scholarly aspiration levels.

2. Much of this is taken from Bendor and Hammond 1992, pp. 311–13.

3. I was so struck by the discrepancy that I e-mailed Simon about the matter. Although he didn't know me and must have received hundreds of e-mail messages weekly from total strangers, Simon was kind enough to reply. He said, "It has been a long time since I looked at Graham Allison's book, *Essence of Decision,* and I shall have to go back and see what he says. But you are wholly right in accusing him, on the basis of your description of his analysis, of misinterpreting the "Carnegie School" views. Perhaps my loudest paen on organizations is one that even predates the work here at CMU—see pp. 100–102 of *Administrative Behavior"* (e-mail, October 24, 1995).

4. See Newell 1990, pp. 121–23, for a cogent discussion of different time scales for information processing.

5. Allison mentions this advantage once (1971, p. 80). He does not, however, emphasize it.

6. For readers who teach an undergraduate course on bureaucracy, the following is an interesting exercise. At the start of the first session, ask your students to free associate in response to the word *bureaucracy.* I guarantee that most of the words they generate will be negative. "Routines" or near synonyms (standard operating procedures, etc.) will usually be mentioned.

7. "If the SOPs are appropriate, average performance—i.e., performance averaged over the range of cases—is better than it would be if each instance were approached individually (given fixed talent, timing, and resource constraints)" (Allison 1971, p. 89). He immediately adds, however, "But specific instances, particularly critical instances that typically do not have 'standard' characteristics, are often handled sluggishly or inappropriately" (p. 89).

8. In a striking essay, Norton Long also differed sharply with the viewpoint of *Essence of Decision,* arguing that the *bureaucracy,* via its "fact-finding [and] policy proposal . . . procedures," must discipline the tendency of *politicians* to manipulate "a free-wheeling world of rhetoric and emotion, . . . a world in which solid facts evaporate" (1954, p. 28). Long's remarks are eerily relevant to the Bush administration's planning for and conduct of the Iraq war.

9. Cognitive diversity and parallel processing can together yield surprisingly powerful effects. For a penetrating, book-length analysis of the advantages of this combination for problem solving, see Page 2007a; for a crisp summary, see Page 2007b.

10. Given the high status enjoyed by most professions in modern societies, it is entirely predictable that some groups will try to pass themselves off as professions even though they lack the key property of codified knowledge. The existence of bogus professions should not blind us to the reality of real ones.

Postmodernist doubts of this kind can be easily dispelled by, for example, contemplating brain surgery executed (pun intended) by someone who lacks the requisite training.

11. However, per Simon's definition of "decision," to see these as value-free choices or purely technical matters involves a fundamental misconceptualization. *All* choices are selections of alternatives based on both factual *and* valuational premises (1947, pp. 4–8; 1952). The choice of what grade cement to use in an ordinary highway is guided by goal premises just as is the selection of Zyklon B for gas chambers. That the former is a mundane matter (at least for clean governments) while the latter is a monstrosity shouldn't obscure important choice-theoretic similarities.

12. For readers who claim to be complete skeptics, I reoffer the challenge about brain surgery noted above.

13. I have benefited from conversations with Josh Bendor on this topic.

14. Social scientists have long recognized that modernity should be considered a scale, not a binary category. As with the hardness of minerals, we say that organization A is more modern than B. Despite some famous attempts at ideal-type analysis, it is unclear to me what a completely modern organization would look like. Fortunately, we don't need to know that in order to make meaningful ordinal comparisons. Diamonds are harder than emeralds; whether diamonds are at the end of the scale is another matter.

15. The latter is often called *the* scientific method, but some philosophers of science doubt the existence of a unique, unchanging scientific method. See Shapere 1984 for an illuminating discussion.

16. As Udy's studies are old, I asked three eminent organizational sociologists whether they knew of any recent empirical studies. They didn't. Weber's ideas seem to have fallen out of favor in organization theory.

17. This seems to be true whether an organization generates useful ideas in-house or borrows them. Indeed, the two capacities appear to go together: the more an organization already knows, the more it can absorb (learn) from other organizations (Cohen and Levinthal 1990).

18. Estimates vary widely. For example, Van der Panne, van Beers, and Kleinknecht (2003, p. 2) report a success rate of 20 percent; Roberts and Brandenburg state that the rate of technical success was over 44 percent (cited in Scherer 1970, p. 355). Part of the difference turns on when a project was considered to be initiated—the further back in R&D, the higher the failure rate—and whether only technical success or overall commercial viability was the criterion of interest.

19. Firms are the other obvious choice, as Weber and Weberians such as Stinchcombe realized.

20. Evidently, however, "soft" aspects of a modern bureaucracy—administration and personnel systems—were modernizing more slowly than hard technology was (Woodward 1965, p. 164).

21. For an insightful analysis of what enables organizations to learn from each other—their "absorptive capacity"—see Cohen and Levinthal 1990.

22. I thank Jacob Shapiro for researching this topic.

23. Apparently ascriptive barriers were higher in the Russian navy (Woodward 1965, p. 126).

24. A *scale* of modernity is still further apart from hypotheses about the dynamics of real systems as measured by that scale. A scale does not imply a unidirectional dynamic; indeed, it implies no specific dynamic at all.

25. A few R&D labs, such as Xerox PARC and Bell Labs, may have also enjoyed these advantages.

26. Nisbett and Ross (1980) provide a beautiful exception to this pattern. A reader of an early draft wrote, "If we're so dumb, how'd we get to the moon?" The authors' reply, in their concluding chapter, was startlingly organizational.

27. The word *ameliorate* is crucial. As Meehl (1992, pp. 351, 372) pointed out, scientific procedures and institutions needn't be perfect; they need only to generate knowledge faster than all other nonscientific procedures and institutions do.

28. However, some of what we think we know may be wrong. For example, the empirical support for the group-think hypothesis, vis à vis Janis's original contexts (the Bay of Pigs and Vietnam), is weaker than commonly believed (Kramer 1998). More generally, consider two hypotheses. Any claim that is consistent with the above cultural fixation will be easily accepted, including any claim with slender empirical support. Inconsistent claims will be assimilated more slowly or, as in Allison's case, misunderstood.

29. Organization theorists and more recently economists have intensively examined the problems of communication flowing upward in hierarchies. Wilensky 1967 is a classic.

30. Rational choice theorists who are purists might study this phenomenon via an incentives-and-incomplete-information model with fully rational decision makers. Some of the implications of such models will overlap with those of bounded rationality models of dictators. I suspect, however, that there will be testable differences. For example, a model of a cognitively constrained dictator might imply overconfidence on the dictator's part. There are many things the tyrant doesn't know but doesn't know that he or she doesn't know. In a canonical rational choice model, the dictator would have this second-order knowledge. Of course, a generous use of ad hoc assumptions about preferences and/or initial beliefs could probably make all testable differences vanish. Such moves are scientifically naughty but very tempting: nobody likes being demonstrably wrong.

31. See Scott 1998 for a penetrating analysis.

32. Stalin's incapacitation was an instance of hot cognition: stress-related impairment. This book emphasizes problems caused by cold cognition, but obviously the hot kind is also important.

33. See Page 2007a for an enlightening and wide-ranging discussion of diversity. His book discusses hiring and several other organizational procedures that can reduce diversity and independence.

34. There is evidence (Kunda and Nisbett 1986) that we overestimate how well we can predict a person's behavior at $t+1$ based on an observation at $t$. If this holds specifically for performance-related behavior, then it implies that we overestimate how tightly correlated a person's performance is over time. Properties such as skill or competence are not mechanically evoked: they require diligence, effort, and focused attention. None of these is automatic, as every good sports coach knows. This suggests that we overestimate the temporal stability of fuzzier variables such as good leadership—or, more magically, charisma.

35. One could say that no one has special expertise on value premises, but moral philosophers might quibble with that. They have expertise in reasoning coherently about value premises but not about the content.

36. A rough guess: if in year $t$ a critic of rational choice theorizing identifies an empirical domain as being completely resistant to such theorizing, by year $t + 10$ some clever economist or game theorist will have produced a formal model, complete with implications, in that domain.

37. Some scholars who work only in the rational choice program may disagree with this last claim, arguing that decision makers learn to optimize in the long run. This is a perfectly understandable defense of the incumbent program's hard core, but I think it is flat wrong. (Indeed, it was partly to disabuse colleagues of the conjecture that the best-known heuristic, satisficing, converges to optimizing in the long run that Kumar, Siegel, and I wrote our 2004 paper on satisficing.) For a cogent critique of this and related rebuttals, see Conlisk 1996.

38. A research program's hard core is often built slowly and painstakingly; nor is it given up lightly. Economists have vigorously–and quite appropriately—defended the hard core of the optimization program. The details of this history are quite interesting to historians and philosophers of science (Hausman 1992 provides a good account), but the intensity of the defense should not surprise anyone familiar with the dynamics of research programs. Experiencing that resistance can be frustrating for young turks—including, of course, the young Herbert Simon—but it is a typical feature of a research program's dynamics.

# References

Abelson, Robert, and Ariel Levi. 1985. "Decision Making and Decision Theory." In G. Lindzey and E. Aronson (eds.), *The Handbook of Social Psychology* 1:231–309. New York: Random House.

Allison, Graham. 1969. "Conceptual Models and the Cuban Missile Crisis." *American Political Science Review* 63:689–718.

Allison, Graham. 1971. *Essence of Decision.* Boston: Little, Brown.

Anderson, John. 1995. *Cognitive Psychology and Its Implications.* 4th ed. New York: W. H. Freeman and Company.

Anderson, Paul, and Gregory Fischer. 1986. "A Monte Carlo Model of a Garbage Can Decision Process." In March and Weissinger-Baylon (eds.), *Ambiguity and Command*, pp. 140–64.

Argote, Linda, and Dennis Epple. 1990. "Learning Curves in Manufacturing." *Science* 247:920–25.

Arrow, Kenneth. 1964. "Review of *A Strategy of Decision,* by David Braybrooke and Charles Lindblom." *Political Science Quarterly* 79:584–88.

Arrow, Kenneth. 1974. *The Limits of Organization.* New York: Norton.

Art, Robert. 1973. "Bureaucratic Politics and American Foreign Policy: A Critique." *Policy Sciences* 4:467–90.

Austen-Smith, David, and Jeffrey Banks. 1996. "Information Aggregation, Rationality and the Condorcet Jury Theorem." *American Political Science Review* 90(1):34–45.

Axelrod, Robert. 1976. *Structure of Decision.* Princeton, NJ: Princeton University Press.

Baguley, Thom, and S. Ian Robertson. 2000. "Where Does Fast and Frugal Cognition Stop? The Boundary between Complex Cognition and Simple Heuristics." *Behavioral and Brain Sciences* 23:742–43.

Behn, Robert, and James Vaupel. 1982. *Quick Analysis for Busy Decision Makers.* New York: Basic Books.

Bendor, Jonathan. 1985. *Parallel Systems: Redundancy in Government.* Berkeley: University of California Press.

Bendor, Jonathan. 1995. "A Model of Muddling Through."*American Political Science Review* 89(4):819–40.

Bendor, Jonathan. 2001. "Bounded Rationality in Political Science." In N. Smelser and P. Baltes (eds.), *International Encyclopedia of the Social & Behavioral Sciences*, pp. 1303–7. Amsterdam: Elsevier.

Bendor, Jonathan, Daniel Diermeier, and Michael Ting. 2003a. "A Behavioral Model of Turnout." *American Political Science Review:* 261–80.

Bendor, Jonathan, Daniel Diermeier, and Michael Ting. 2003b. "The Empirical Content of Behavioral Models of Adaptation." Presented at the Annual Meeting of the Midwest Political Science Association, Chicago, IL, April 3–6.

Bendor, Jonathan, Daniel Diermeier, and Michael Ting. 2007. "Comment: Adaptive Models in Sociology and the Problem of Empirical Content." *American Journal of Sociology* 112(5):1534–45.

Bendor, Jonathan, and Thomas Hammond. 1992. "Rethinking Allison's Models." *American Political Science Review* 86:301–22.

Bendor, Jonathan, and Sunil Kumar. 2005. "The Perfect is the Enemy of the Best: Adaptive versus Optimal Organizational Reliability." *Journal of Theoretical Politics* 17(1):5–39.

Bendor, Jonathan, Sunil Kumar, and David Siegel. 2004. "Satisficing: A *Pretty* Good Heuristic." Presented at the Annual Meeting of the Midwest Political Science Association, Chicago, IL, April 15–18.

Bendor, Jonathan, Sunil Kumar, and David Siegel. 2009. "Satisficing: A 'Pretty Good' Heuristic." *Berkeley Electronic Journal of Theoretical Economics* (Advances) 9 (1) article 9.

Bendor, Jonathan, and Terry Moe. 1986. "Agenda Control, Committee Capture, and the Dynamics of Institutional Politics." *American Political Science Review* 80:1187–1207.

Bendor, Jonathan, Terry M. Moe, and Kenneth W. Shotts. 1996. "Recycling the Garbage Can: An Assessment of the Research Program." Graduate School of Business, Stanford University. Typescript.

Bendor, Jonathan, Terry Moe, and Kenneth Shotts. 2001. "Recycling the Garbage Can: An Assessment of the Research Program." *American Political Science Review* 95(1):169–90.

Bendor, Jonathan, Dilip Mookherjee, and Debraj Ray. 2006. "Satisficing and Selection in Electoral Competition." *Quarterly Journal of Political Science* 1(2):171–200.

Bettman, James, Mary Frances Luce, and John Payne. 1998. "Constructive Consumer Choice Processes." *Journal of Consumer Research* 25 (December): 187–217.

Black, Max. 1962. *Models and Metaphors*. Ithaca, NY: Cornell University Press.

Boland, Lawrence. 1979. "A Critique of Friedman's Critics." *Journal of Economic Literature* 17:503–22.

Boulding, Kenneth. 1964. "Review of *A Strategy of Decision*, by David Braybrooke and Charles Lindblom." *American Sociological Review* 29:930–31.

Braybrooke, David. 1985. "Scale, Combination, Opposition—A Rethinking of Incrementalism." *Ethics* 95:920–33.

Braybrooke, David, and Charles Lindblom. 1963. *A Strategy of Decision*. New York: Free Press.

Bush, Robert, and Frederick Mosteller. 1955. *Stochastic Models for Learning*. New York: Wiley.

Byron, Michael (ed.). 2004. *Satisficing and Maximizing: Moral Theorists on Practical Reason*. Cambridge: Cambridge University Press.

Camerer, Colin. 1994. "Individual Decision Making." In J. Kagel and A. Roth (eds.), *Handbook of Experimental Economics*, pp. 587–703. Princeton, NJ: Princeton University Press.

Camerer, Colin. 2003. *Behavioral Game Theory*. Princeton, NJ: Princeton University Press.

Camerer, Colin, and Eric Johnson. 1991. "The Process-Performance Paradox in Expert Judgment: How Can Experts Know So Much and Predict So Badly?" In A. Ericsson and J. Smith (eds.), *Toward a General Theory of Expertise: Prospects and Limits*, pp. 195–217. Cambridge: Cambridge University Press.

Carley, Kathleen. 1986. "Measuring Efficiency in a Garbage Can Hierarchy." In March and Weissinger-Baylon (eds.), *Ambiguity and Command*, pp. 165–94.

Carpenter, Daniel, and Michael Ting. 2007. "Regulatory Errors with Endogenous Agendas." *American Journal of Political Science* 51(4):835–52.

Chase, Valerie, Ralph Hertwig, and Gerd Gigerenzer. 1998. "Visions of Rationality." *Trends in Cognitive Science* 2:206–14.

Christensen-Szalanski, Jay, and Lee Beach. 1984. "The Citation Bias: Fad and Fashion in the Judgment and Decision Literature." *American Psychologist* 39: 75–78.

Cohen, Michael, and James March. 1986. *Leadership and Ambiguity: The American College President*. 2nd ed. Boston: Harvard Business School Press.

Cohen, Michael, James March, and Johan Olsen. 1972. "A Garbage Can Model of Organizational Choice." *Administrative Science Quarterly* 17 (March): 1–25.

Cohen, Michael, James March, and Johan Olsen. 1976. "People, Problems, Solutions and the Ambiguity of Relevance." In March and Olsen (eds.), *Ambiguity and Choice in Organizations*, pp. 24–37.

Cohen, Wesley, and Daniel Levinthal. 1990. "Absorptive Capacity: A New Perspective on Learning and Innovation." *Administrative Science Quarterly* 35: 128–52.

Conlisk, John. 1996. "Why Bounded Rationality?" *Journal of Economic Literature* 34:669–700.

Converse, Phillip. 1975. "Public Opinion and Voting Behavior." In F. Greenstein and N. Polsby (eds.), *Handbook of Political Science* 4:75–169. Reading, MA: Addison-Wesley.

Cooper, Richard. 2000. "Simple Heuristics Could Make Us Smart, But Which Heuristics Do We Apply When?" *Behavioral and Brain Sciences* 23:746.

Cosmides, Leda, and John Tooby. 1992. "Cognitive Adaptations for Social Exchange." In J. Barkow, L. Cosmides, and J. Tooby (eds.), *The Adapted Mind: Evolutionary Psychology and the Generation of Culture*, pp. 163–228. New York: Oxford University Press.

Cowan, Nelson. 2001. "The Magical Number 4 in Short-term Memory: A Reconsideration of Mental Storage Capacity." *Behavioral and Brain Sciences* 24:87–185.

Crecine, John. 1969. *Governmental Problem-Solving.* Chicago: Rand McNally.

Crecine, John. 1970. "Defense Budgeting: Organizational Adaptation to Environmental Constraints." Monograph RM-6121–PR. Santa Monica: Rand Corporation.

Crecine, John. 1985. "A Positive Theory of Public Spending." In L. Sproull and P. Larkey (eds.), *Advances in Information Processing in Organizations* 2:99–154. Greenwich, CT: JAI Press.

Crecine, Patrick. 1986. "Defense Resource Allocation." In March and Weissinger-Baylon (eds.), *Ambiguity and Command*, pp. 72–119.

Cyert, Richard, and James March. 1963. *A Behavioral Theory of the Firm.* Englewood Cliffs, NJ: Prentice Hall.

Cyert, Richard, and James March. 1992. Epilogue to *A Behavioral Theory of the Firm*, pp. 214–46. Oxford: Blackwell.

Dahl, Robert, and Charles Lindblom. 1953. *Politics, Economics, and Welfare.* New York: Harper & Row.

Davis, Otto, Michael Dempster, and Aaron Wildavsky. 1966. "A Theory of the Budget Process." *American Political Science Review* 60:529–47.

Davis, Otto, Michael Dempster, and Aaron Wildavsky. 1974. "Towards a Predictive Theory of Government Expenditure." *British Journal of Political Science* 4:419–52.

Dawes, Robyn. 1998. "Behavioral Decision Making and Judgment." In D. Gilbert, S. Fiske, and G. Lindzey (eds.), *The Handbook of Social Psychology* 1:497–548. Boston: McGraw-Hill.

Dawes, Robyn, D. Faust, and Paul Meehl. 1989. "Clinical versus Actuarial Judgment." *Science* 243:1668–74.

Derthick, Martha. 1990. *Agency under Stress.* Washington, DC: Brookings Institution.

Dewey, John. 1927. *The Public and Its Problems.* Denver: Henry Holt.

Diener, Ed, and Eunkook Suh. 1999. "National Differences in Subjective Well-Being." In D. Kahneman, E. Diener, and N. Schwarz (eds.), *Well-Being: The Foundations of Hedonic Psychology*, pp. 434–50. New York: Russell Sage.

Diener, Ed, Eunkook Suh, Richard Lucas, and Heidi Smith. 1999. "Subjective Well-Being: Three Decades of Progress." *Psychological Bulletin* 75(2):276–302.

Downs, Anthony. 1957. *An Economic Theory of Democracy*. New York: Harper.

Dror, Yehezkiel. 1964. "Muddling Through—'Science' or Inertia?" *Public Administration Review* 24:153–57.

Dubnick, Melvin. 2002. "Once a Political Scientist, Always . . ." In *Remembering a Giant: A Tribute to Herbert A. Simon*. Washington, DC: American Political Science Association www.apsanet.org/new/simon.

Einhorn, Hillel, and Robin Hogarth. 1981. "Behavioral Decision Theory." In M. Rosenzweig and L. Porter (eds.), *Annual Review of Psychology* 32:53–88. Palo Alto: Annual Reviews.

Epstein, Steven. 1997. "Activism, Drug Regulation, and the Politics of Therapeutic Evaluation in the AIDS Era." *Social Studies of Science* 27(5):691–726.

Erdfelder, Edgar, and Martin Brandt. 2000. "How Good Are Fast and Frugal Inference Heuristics in Case of Limited Knowledge?" *Behavioral and Brain Sciences* 23:747–49.

Ericsson, Anders, and A. Lehmann. 1996. "Expert and Exceptional Performance: Evidence on Maximal Adaptations on Task Constraints." *Annual Review of Psychology* 47:273–305.

Etzioni, Amitai. 1967. "Mixed Scanning: A 'Third' Approach to Decision-making." *Public Administration Review* 27:385–92.

Evans, David, and Mark Peattie. 1997. *Kaigun: Strategy, Tactics, and Technology in the Imperial Japanese Navy, 1887–1941*. Annapolis, MD: U.S. Naval Institute Press.

Evans, Jonathan, and David Over. 1996. *Rationality and Reasoning*. Hove, Sussex: Psychology Press.

Ferejohn, John, Morris Fiorina, and Edward Packel. 1980. "Nonequilibrium Solutions for Legislative Systems." *Behavioral Science* 25:140–48.

Ferejohn, John, Richard McKelvey, and Edward Packel. 1984. "Limiting Distributions for Continuous State Markov Voting Models." *Social Choice and Welfare* 1:45–67.

Fiorina, Morris. 1990. "Information and Rationality in Elections." In J. Ferejohn and J. Kuklinski (eds.), *Information and Democratic Processes*, pp. 329–42. Champaign: University of Illinois Press.

Friedman, Jeffrey. 1996. *The Rational Choice Controversy: Economic Models of Politics Reconsidered*. New Haven, CT: Yale University Press.

Friedman, Milton. 1953. "The Methodology of Positive Economics." In *Essays in Positive Economics*, pp. 3–43. Chicago: University of Chicago Press.

Gallucci, Robert. 1975. *Neither Peace nor Honor: The Politics of American Military Policy in Vietnam*. Baltimore: Johns Hopkins University Press.

Gawande, Atul. 2002. *Complications: A Surgeon's Notes on an Imperfect Science*. New York: Henry Holt.

Gerth, Hans, and C. Wright Mills. 1946. *From Max Weber*. Oxford: Oxford University Press.

Gigerenzer, Gerd. 1991. "How to Make Cognitive Illusions Disappear: Beyond Heuristics and Biases." *European Review of Social Psychology* 2:83–115.

Gigerenzer, Gerd. 2001. "The Adaptive Toolbox." In Gigerenzer and Selten (eds.), *Bounded Rationality*, pp. 37–50.

Gigerenzer, Gerd. 2004. "Striking a Blow for Sanity in Theories of Rationality." In M. Augier and J. March (eds.), *Models of a Man: Essays in Memory of Herbert A. Simon*, pp. 389–410. Cambridge, MA: MIT Press.

Gigerenzer, Gerd, Jean Czerlinski, and Laura Martignon. 1999. "How Good Are Fast and Frugal Heuristics?" In J. Shanteau, B. Mellers, and D. Schum (eds.), *Decision Science and Technology: Reflections on the Contributions of Ward Edwards*, pp. 81–103. Boston: Kluwer.

Gigerenzer, Gerd, and Daniel Goldstein. 1996. "Reasoning the Fast and Frugal Way: Models of Bounded Rationality." *Psychological Review* 103:650–69.

Gigerenzer, Gerd, and Reinhard Selten (eds.) 2001. *Bounded Rationality: The Adaptive Toolbox*. Cambridge, MA: MIT Press.

Gigerenzer, Gerd, P. Todd, and the ABC Group (eds.) 1999. *Simple Heuristics That Make Us Smart*. Oxford: Oxford University Press.

Gilbert, Daniel. 1998. "Ordinary Personology." In D. Gilbert, S. Fiske, and G. Lindzey (eds.), *The Handbook of Social Psychology* 2:89–150. Boston: McGraw-Hill.

Gilovich, Thomas, and Dale Griffin. 2002. "Introduction—Heuristics and Biases: Then and Now." In T. Gilovich, D. Griffin, and D. Kahneman (eds.), *Heuristics and Biases: The Psychology of Intuitive Judgment*, pp. 1–18. Cambridge: Cambridge University Press.

Gilovich, Thomas, Dale Griffin, and Daniel Kahneman (eds.). 2002. *Heuristics and Biases: The Psychology of Inuitive Judgment*. Cambridge: Cambridge University Press.

Gode, Dhananjay, and Shyam Sunder. 1993. "Allocative Efficiency of Markets with Zero-Intelligence Traders: Markets as a Partial Substitute for Individual Rationality." *Journal of Political Economy* 101:119–37.

Goldstein, Daniel, Gerd Gigerenzer, Robin Hogarth, Alex Kacelnik, Yaakov Kareev, Gary Klein, Laura Martignon, John Payne, and Karl Schlag. 2001. "Group Report: Why and When Do Simple Heuristics Work?" In Gigerenzer and Selten (eds.), *Bounded Rationality*, pp. 173–90.

Goodin, Robert. 1999. "Rationality Redux: Reflections on Herbert A. Simon's Vision of Politics." In J. Alt, M. Levi, and E. Ostrom (eds.), *Competition and Cooperation: Conversations with Nobelists about Economics and Political Science*, pp. 60–84. New York: Russell Sage.

Goodin, Robert, and Hans-Dieter Klingemann. 1996. "Political Science: The Discipline." In Goodin and Klingemann (eds.), *A New Handbook of Political Science*, pp. 3–49. New York: Oxford University Press.

Goodin, Robert, and Ilmar Waldner. 1979. "Thinking Big, Thinking Small, and Not Thinking at All." *Public Policy* 27:1–24.

Gopnik, Allison, Andrew Meltzoff, and Patricia Kuhl. 1999. *The Scientist in the Crib: Minds, Brains, and How Children Learn*. New York: William Morrow.

Green, Donald, and Ian Shapiro. 1994. *Pathologies of Rational Choice Theory: A Critique of Applications in Political Science*. New Haven, CT: Yale University Press.

Green, Mark, and Frank Thompson. 2001. "Organizational Process Models of Budgeting." In J. Bartle (ed.), *Evolving Theories of Public Budgeting*, pp. 55–81. Amsterdam: Elsevier Science.

Griffin, Dale, Richard Gonzalez, and Carol Varey. 2001. "The Heuristics and Biases Approach to Judgment Under Uncertainty." In A. Tesser and N. Schwarz (eds.), *Blackwell Handbook of Social Psychology: Intraindividual Processes*, pp. 207–35. Malden, MA: Blackwell.

Griffin, Dale, and Daniel Kahneman. 2003. "Judgmental Heuristics: Human Strengths or Human Weaknesses." In L. Aspinwall and U. Staudinger (eds.), *A Psychology of Human Strengths: Perspectives on an Emerging Field*, pp. 165–78. New York: APA Books.

Grofman, Bernard, and Scott Feld. 1988. "Rousseau's General Will: A Condorcetian Perspective." *American Political Science Review* 82:567–76.

Grofman, Bernard, Guillermo Owen, and Scott Feld. 1983. "Thirteen Theorems in Search of the Truth." *Theory and Decision* 15:261–78.

Guarnaschelli, Serena, Richard McKelvey, and Thomas Palfrey. 2000. "An Experimental Study of Jury Decision Rules." *American Political Science Review* 94:407–24.

Gulick, Luther. 1937. "Notes on the Theory of Organization." In L. Gulick and L. Urwick (eds.), *Papers on the Science of Administration*. New York: Columbia University.

Hammond, Kenneth. 1990. "Functionalism and Illusionism: Can Integration Be Usefully Achieved?" In R. Hogarth (ed.), *Insight in Decision Making*, pp. 227–61. Chicago: University of Chicago Press.

Hammond, Thomas. 2002. "Heuristic Search and the Power of Hierarchy." Presented at the Herbert A. Simon Award ceremony, Midwest Political Science Association Meetings, Chicago, IL.

Hand, Learned. 1960. *The Spirit of Liberty: Papers and Addresses of Learned Hand*. New York: Knopf.

Hausman, Daniel. 1992. *The Inexact and Separate Science of Economics*. Cambridge: Cambridge University Press.

Heimann, C. F. Larry. 1993. "Understanding the *Challenger* Disaster: Organizational Structure and the Design of Reliable Systems." *American Political Science Review* 87:421–35.

Heimann, C. F. Larry. 1997. *Acceptable Risks: Politics, Policy, and Risky Technologies*. Ann Arbor: University of Michigan Press.

Heimer, Carol, and Arthur Stinchcombe. 1998. "Remodeling the Garbage Can." Northwestern University. Typescript.

Hertwig, Ralph, and Andreas Ortmann. 2005. "The Cognitive Illusion Controversy: Why This Methodological Debate in Disguise Should Matter to Economists." In Rami Zwick and Amnon Rapoport (eds.), *Experimental Business Research* 3:113–30. Boston: Kluwer.

Hertwig, Ralph, and Peter Todd. 2003. "More is Not Always Better: The Benefits of Cognitive Limits." In D. Hardman and L. Macchi (eds.), *Thinking: Psychological Perspectives on Reasoning, Judgment and Decision Making*, pp. 213–31. Chichester: Wiley.

Hilgard, Ernest, and Gordon Bower. 1966. *Theories of Learning.* 3rd ed. New York: Appleton-Century-Crofts.

Hogarth, Robin. 1987. *Judgment and Choice.* 2nd ed. New York: Wiley.

Holyoak, Keith. 1990. "Problem Solving." In D. Osherson and E. Smith (eds.), *An Invitation to Cognitive Science* 3:117–46. Cambridge, MA: MIT Press.

Huberman, Bernardo, and Natalie Glance. 1993. "Evolutionary Games and Computer Simulations." *Proceedings of the National Academy of Science* 90 (August): 7716–8.

Hurley, Susan, and Nick Chater (eds.). 2005. *Perspectives on Imitation: From Neuroscience to Social Science.* Vol. 2. Cambridge, MA: MIT Press.

Jane, Fred. 1904. *The Japanese Navy.* London: Thacker.

Jefferies, Chris. 1977. "Defense Decisionmaking in the Organizational-Bureaucratic Context." In J. Endicott and R. Stafford Jr. (eds.), *American Defense Policy,* pp. 227–39. 4th ed. Baltimore: Johns Hopkins University Press.

Jervis, Robert. 1976. *Perception and Misperception in International Politics.* Princeton, NJ: Princeton University Press.

Jones, Bryan. 1999. "Bounded Rationality." In N. Polsby (ed.), *Annual Review of Political Science* 2:297–321. Palo Alto: Annual Reviews.

Jones, Bryan. 2001. *Politics and the Architecture of Choice.* Chicago: University of Chicago Press.

Kahneman, Daniel. 2003. "A Perspective on Judgment and Choice: Mapping Bounded Rationality." *American Psychologist* 58(9): 697–720.

Kahneman, Daniel, Ed Diener, and Nobert Schwarz (eds.). 1999. *Well-Being: The Foundations of Hedonic Psychology.* New York: Russell Sage.

Kahneman, Daniel, and Shane Frederick. 2002. "Representativeness Revisited: Attribute Substitution in Intuitive Judgment." In T. Gilovich, D. Giffin, and D. Kahneman (eds.), *Heuristics and Biases: The Psychology of Intuitive Judgment,* pp. 49–81. Cambridge: Cambridge University Press.

Kahneman, Daniel, Jack Knetsch, and Richard Thaler. 1990. "Experimental Tests of the Endowment Effect and the Coase Theorem." *Journal of Political Economy* 98:1325–48.

Kahneman, Daniel, and Alan Krueger. 2006. "Developments in the Measurement of Subjective Well-Being." *Journal of Economic Perspectives* 20(1):3–24.

Kahneman, Daniel, Paul Slovic, and Amos Tversky. 1982. *Judgment under Uncertainty: Heuristics and Biases.* Cambridge: Cambridge University Press.

Kahneman, Daniel, and Amos Tversky. 1979. "Prospect Theory: An Analysis of Decision under Risk." *Econometrica* 47:263–91.

Kahneman, Daniel, and Amos Tversky. 1996. "On the Reality of Cognitive Illusions: A Reply to Gigerenzer's Critique." *Psychological Review* 103:582–91.

Karandikar, Rajeeva, Dilip Mookherjee, Debraj Ray, and Fernando Vega-Redondo. 1998. "Evolving Aspirations and Cooperation." *Journal of Economic Theory* 80: 292–331.

Kassiola, Joel. 1974. "Fallibilism and Political Knowledge." Ph.D. thesis. Department of Politics, Princeton University.

Kemeny, John, and J. Laurie Snell. 1960. *Finite Markov Chains*. Princeton, NJ: Van Nostrand.

Kinder, Donald. 1998. "Opinion and Action in the Realm of Politics." In D. Gilbert, S. Fiske, and G. Lindzey (eds.), *The Handbook of Social Psychology* 2:778–867. Boston: McGraw-Hill.

Kingdon, John. 1984. *Agendas, Alternatives, and Public Policies*. Boston: Little, Brown.

Knott, Jack, and Gary Miller. 1981. "Comment on Lustick." *American Political Science Review* 75:725–27.

Kohn, Meier, and Steven Shavell. 1974. "The Theory of Search." *Journal of Economic Theory* 9:93–123.

Kollman, Ken, John Miller, and Scott Page. 1992. "Adaptive Parties in Spatial Elections." *American Political Science Review* 86:929–37.

Kollman, Ken, John Miller, and Scott Page. 1998. "Political Parties and Electoral Landscapes." *British Journal of Political Science* 28:139–58.

Konrad, Alison, and Jeffrey Pfeffer. 1990. "Do You Get What You Deserve? Factors Affecting the Relationship between Productivity and Pay." *Administrative Science Quarterly* 35:258–85.

Kotovsky, Kenneth, John Hayes, and Herbert Simon. 1985. "Why Are Some Problems Hard?" *Cognitive Psychology* 17:248–94.

Kramer, Roderick. 1998. "Revisiting the Bay of Pigs and Vietnam Decisions 25 Years Later: How Well Has the Groupthink Hypothesis Stood the Test of Time?" *Organizational Behavior and Human Decision Processes* 73:236–71.

Krasner, Steven. 1972. "Are Bureaucracies Important? (Or Allison Wonderland)." *Foreign Policy* 7:159–79.

Kuklinski, James, and Paul Quirk. 2000. "Reconsidering the Rational Public: Cognition, Heuristics, and Mass Opinion." In A. Lupia, M. McCubbins, and S. Popkin (eds.), *Elements of Reason*, pp. 153–82. New York: Cambridge University Press.

Kunda, Ziva, and Richard Nisbett. 1986. "Prediction and the Partial Understanding of the Law of Large Numbers." *Journal of Experimental Social Psychology* 22(4):339–54.

Ladha, Krishna. 1992. "The Condorcet Jury Theorem, Free Speech, and Correlated Votes." *American Journal of Political Science* 36:597–634.

Lakatos, Imre. 1970. "Falsification and the Methodology of Scientific Research Programmes." In I. Lakatos and A. Musgrave (eds.), *Criticism and the Growth of Knowledge*, pp. 91–196. Cambridge: Cambridge University Press.

Landau, Martin. 1969. "Redundancy, Rationality, and the Problem of Duplication and Overlap." *Public Administration Review* 29:346–58.

Landau, Martin. 1972. *Political Science and Political Theory: Studies in the Methodology of Political Inquiry*. New York: Macmillan.

Landau, Martin. 1973. "On the Concept of a Self-Correcting Organization." *Public Administration Review* 33(6):533–42.

Landau, Martin. 1977. "The Proper Domain of Policy Analysis." *American Journal of Political Science* 21:423–27.

Landau, Martin, and Donald Chisholm. 1995. "The Arrogance of Optimism: Notes on Failure-Avoidance Management." *Journal of Contingencies and Crisis Management* 3:67–80.

Landes, David. 1999. *The Wealth and Poverty of Nations: Why Some Are So Rich and Some So Poor.* New York: Norton.

LaPorte, Todd, and Paula Consolini. 1991. "Working in Practice But Not in Theory: Theoretical Challenges of 'High-Reliability Organizations.'" *Journal of Public Administration Research and Theory* 1:19–47.

La Rochefoucauld, Duc de. 1678. Maxims. New York: Penguin Classics.

Lau, Richard, and David Redlawsk. 2001. "Advantages and Disadvantages of Cognitive Heuristics in Political Decision Making." *American Journal of Political Science* 45(4):951–71.

Laudan, Rachel, Larry Laudan, and Arthur Donovan. 1988. "Testing Theories of Scientific Change." In A. Donovan, L. Laudan, and R. Laudan (eds.), *Scrutinizing Science*, pp. 3–44. Dordrecht: Kluwer.

Layard, Richard. 2006. "Happiness and Public Policy: A Challenge to the Profession." *Economic Journal* 116:C24–C33.

Levy, Jack. 1986. "Organizational Routines and the Causes of War." *International Studies Quarterly* 30:193–222.

Lindblom, Charles. 1959. "The Science of 'Muddling Through.'" *Public Administration Review* 19:79–88.

Lindblom, Charles. 1964. "Contexts for Change and Strategy: A Reply." *Public Administration Review* 24:157–58.

Lindblom, Charles. 1965. *The Intelligence of Democracy.* New York: Free Press.

Lindblom, Charles. 1979. "Still Muddling, Not Yet Through." *Public Administration Review* 39:517–26.

Linden, David. 2007. *The Accidental Mind: How Brain Evolution Has Given Us Love, Memory, Dreams, and God.* Cambridge, MA: Belknap Press of Harvard University Press.

Lippmann, Walter. 1922. *Public Opinion.* New York: Macmillan.

List, Christian, and Robert Goodin. 2001. "Epistemic Democracy: Generalizing the Condorcet Jury Theorem." *Journal of Political Philosophy* 9:277–306.

Lodge, Milton. 1995. "Toward a Procedural Model of Candidate Evaluation." In M. Lodge and K. McGraw (eds.), *Political Judgment*, pp. 111–40. Ann Arbor: University of Michigan Press.

Lohmann, Susanne, and Hugo Hopenhayn. 1998. "Delegation and the Regulation of Risk." *Games and Economic Behavior* 23:222–46.

Long, Norton. 1954. "Public Policy and Administration: The Goals of Rationality and Responsibility." *Public Administration Review* 14:22–31.

Lopes, Lola. 1981. "Performing Competently." *Behavioral and Brain Sciences* 4:343–44.

Luce, R. Duncan. 2000. "Fast, Frugal, and Surprisingly Accurate Heuristics." *Behavioral and Brain Sciences* 23:757–58.

Lupia, Arthur. 1994. "Shortcuts versus Encyclopedias: Information and Voting Behavior in California Insurance Reform Elections." American Political Science Review 88(1):63–76.

Lustick, Ian. 1980. "Explaining the Variable Utility of Disjointed Incrementalism: Four Propositions." *American Political Science Review* 74:342–53.

Maddison, Angus. 2005. "Measuring and Interpreting World Economic Performance 1500–2001." *Review of Income and Wealth* Series 51 (1):1–35.

March, James. 1978. "Bounded Rationality, Ambiguity, and the Engineering of Choice." *Bell Journal of Economics* 9 (Autumn): 587–608.

March, James. 1988. "Ambiguity and Accounting: The Elusive Link Between Information and Decision Making." In March, *Decisions and Organizations,* pp. 384–408. New York: Blackwell.

March, James. 1991. "Exploration and Exploitation in Organizational Learning." *Organization Science* 2(1):71–87.

March, James. 1994. *A Primer on Decision Making: How Decisions Happen.* New York: Free Press.

March, James, and Johan Olsen. 1975. "The Uncertainty of the Past: Organizational Learning under Uncertainty." *European Journal of Political Research* 3 (June): 147–71.

March, James, and Johan Olsen (eds.). 1976a. *Ambiguity and Choice in Organizations.* Bergen, Norway: Universitetsforlaget.

March, James, and Johan Olsen. 1976b. "Attention and the Ambiguity of Self-interest." In *Ambiguity and Choice in Organizations,* pp. 38–53.

March, James, and Johan Olsen. 1976c. "Organizational Choice under Ambiguity." In *Ambiguity and Choice in Organizations,* pp. 10–23.

March, James, and Johan Olsen. 1983. "What Administrative Reorganization Tells Us about Governing." *American Political Science Review* 77 (June): 281–96.

March, James, and Johan Olsen. 1984. "The New Institutionalism: Organizational Factors in Political Life." *American Political Science Review* 73 (September): 734–49.

March, James, and Johan Olsen. 1986. "Garbage Can Models of Decision Making in Organizations." In March and Weissinger-Baylon (eds.), *Ambiguity and Command,* pp. 11–35.

March, James, and Johan Olsen. 1989. *Rediscovering Institutions: The Organizational Basis of Politics.* New York: Free Press.

March, James, and Johan Olsen. 1995. *Democratic Governance.* New York: Free Press.

March, James, and Johan Olsen. 1996. "Institutional Perspectives on Political Institutions." *Governance: An International Journal of Policy and Administration* 9 (July): 247–64.

March, James, and Johan Olsen. 2005. "The Institutional Dynamics of International Political Orders." *International Organization* 52:943–69.

March, James, and Pierre Romelaer. 1976. "Position and Presence in the Drift of Decisions." In March and Olsen (eds.), *Ambiguity and Choice in Organizations,* pp. 251–76.

March, James, and Herbert Simon. 1958. *Organizations.* New York: Wiley.

March, James, and Roger Weissinger-Baylon. 1986. *Ambiguity and Command: Organizational Perspectives on Military Decision Making.* Marshfield, MA: Pitman.

Margolis, Howard. 2000. "Simple Heuristics That Make Us Dumb." *Behavioral and Brain Sciences* 23:758.

Masuch, Michael, and Perry LaPotin. 1989. "Beyond Garbage Cans: An AI Model of Organizational Choice." *Administrative Science Quarterly* 34 (March):38–67.

McClellan, Mark. 2007. "Drug Safety Reform at the FDA—Pendulum Swing or Systematic Improvement?" *New England Journal of Medicine* 356(17):1700–2.

McKelvey, Richard. 1976. "Intransitivities in Multidimensional Voting Models and Some Implications for Agenda Control." *Journal of Economic Theory* 12:472–82.

McKelvey, Richard, and Peter Ordeshook. 1990. "Information and Elections: Retrospective Voting and Rational Expectations." In J. Ferejohn and J. Kuklinski (eds.), *Information and Democratic Processes*, pp. 281–312. Champaign: University of Illinois Press.

Meehl, Paul. 1992. "Cliometric Metatheory: The Acturial Approach to Empirical, History-Based Philosophy of Science." *Psychological Reports* 71:339–467.

Mellers, Barbara, Ralph Hertwig, and Daniel Kahneman. 2001. "Do Frequency Representations Eliminate Conjunction Effects? An Exercise in Adversarial Collaboration." *Psychological Science* 12:269–75.

Meyer, John, and Brian Rowan. 1977. "Institutionalized Organizations: Formal Structure as Myth and Ceremony." *American Journal of Sociology* 83 (September): 340–63.

Mezias, Steve, and Mario Scarselletta. 1994. "Resolving Financial Reporting Problems: An Institutional Analysis of the Process." *Administrative Science Quarterly* 39 (December): 654–78.

Miller, George. 1956. "The Magic Number Seven, Plus or Minus Two." *Psychological Review* 63:81–97.

Miller, George. 1989. "Scientists of the Artificial." In D. Klahr and K. Kotovsky (eds.), *Complex Information Processing: The Impact of Herbert A. Simon*, pp. 145–61. Hillsdale, NJ: Lawrence Erlbaum Associates.

Miller, Nicholas. 1986. "Information, Electorates, and Democracy: Some Extensions and Interpretations of the Condorcet Jury Theorem." In B. Grofman and G. Owen (eds.), *Information Pooling and Group Decision Making,* pp. 173–92. Greenwich, CT: JAI Press.

Moch, Michael, and Louis Pondy. 1977. "The Structure of Chaos: Organized Anarchy as a Response to Ambiguity." *Administrative Science Quarterly* 22 (June): 351–62.

Moe, Terry. 1991. "Politics and the Theory of Organization." *Journal of Law, Economics and Organizations* 7:106–29.

Morton, Adam. 2000. "Heuristics All the Way Up?" *Behavioral and Brain Sciences* 23:758.

Narendra, Kumpati, and Mandaym Thathachar. 1989. *Learning Automata: An Introduction.* Englewood Cliffs, NJ: Prentice-Hall.

Neale, Margaret, and Max Bazerman. 1991. *Cognition and Rationality in Negotiations.* New York: Free Press.

Nelson, Richard, and Sidney Winter. 1982. *An Evolutionary Theory of Economic Change.* Cambridge, MA: Belknap Press of Harvard University Press.

Newell, Allen. 1969. "Heuristic Programming: Ill-Structured Problems." In Julius Aronofsky (ed.), *Progress in Operations Research*. Vol. 3. New York: Wiley.

Newell, Allen, and Herbert Simon. 1972. *Human Problem Solving.* Englewood Cliffs, NJ: Prentice-Hall.

Newell, Allen. 1990. *Unified Theories of Cognition.* Cambridge, MA: Harvard University Press.

Nisbett, Richard, and Lee Ross. 1980. *Human Inference: Strategies and Shortcomings of Social Judgment.* Englewood Cliffs, NJ: Prentice-Hall.

Olsen, Johan. 2001. "Garbage Cans, New Institutionalism, and the Study of Politics." *American Political Science Review* 95(1):191–98.

Padgett, John. 1980a. "Bounded Rationality in Budgetary Research." *American Political Science Review* 74:354–72.

Padgett, John. 1980b. "Managing Garbage Can Hierarchies." *Administrative Science Quarterly* 25 (December): 583–604.

Padgett, John. 1981. "Hierarchy and Ecological Control in Federal Budgetary Decision Making." *American Journal of Sociology* 87:75–128.

Page, Scott. 2007a. *The Difference: How the Power of Diversity Creates Better Groups, Firms, Schools, and Societies.* Princeton, NJ: Princeton University Press.

Page, Scott. 2007b. "Making the Difference: Applying a Logic of Diversity." *Academy of Management Perspectives* 21(4):6–20.

Palfrey, Thomas, and Howard Rosenthal. 1985. "Voter Participation and Strategic Uncertainty." *American Political Science Review* 29:62–78.

Parsons, Talcott. 1964. "Evolutionary Universals in Society." *American Sociological Review* 29:339–57.

Payne, John, James Bettman, and Eric Johnson. 1988. "Adaptive Strategy Selection in Decision Making." *Journal of Experimental Psychology: Learning, Memory, and Cognition* 14(3):534–52.

Payne, John, James Bettman, and Eric Johnson. 1993. *The Adaptive Decision Maker.* Cambridge, MA: Cambridge University Press.

Penn, E. Maggie. 2007. "A Model of Farsighted Voting." Harvard University. Unpublished ms.

Perrow, Charles. 1977. "Review of *Ambiguity and Choice.*" *Contemporary Society* 6 (May): 294–98.

Perrow, Charles. 1984. *Normal Accidents: Living with High-Risk Technologies.* New York: Basic Books.

Perrow, Charles. 1986. *Complex Organizations: A Critical Essay.* New York: McGraw-Hill.

Peters, B. Guy. 1996. "Political Institutions, Old and New." In R. Goodin and H. Klingemann (eds.), *A New Handbook of Political Science,* pp. 205–20. New York: Oxford University Press.

Peterson, Paul. 1976. *School Politics, Chicago Style.* Chicago: University of Chicago Press.

Pinker, Steven. 1997. *How the Mind Works.* New York: W. W. Norton.

Platt, John. 1964. "Strong Inference: Certain Systematic Methods of Scientific Thinking May Produce Much More Rapid Progress Than Others." *Science* 146(3642):347–53.

Polya, George. 1945. *How to Solve It*. Princeton, NJ: Princeton University Press.

Popper, Karl. 1963. *Conjectures and Refutations: The Growth of Scientific Knowledge*. New York: Basic Books.

Posen, Barry. 1984. *The Sources of Military Doctrine*. Ithaca, NY: Cornell University Press.

Psillos, Stathis. 2000. "The Present State of the Scientific Realism Debate." *British Journal of the Philosophy of Science* 51:705–28.

Quirk, Paul. 1980. "Food and Drug Administration." In James Q. Wilson (ed.), *The Politics of Regulation*, pp. 191–235. New York: Basic Books.

Richerson, Peter, and Robert Boyd. 2005. *Not by Genes Alone: How Culture Transformed Human Evolution*. Chicago: University of Chicago Press.

Riker, William, and Peter Ordeshook. 1968. "A Theory of the Calculus of Voting." *American Political Science Review* 62:25–42.

Robbins, Herbert, and Sutton Monro. 1951. "A Stochastic Approximation Method." *Annals of Mathematical Statistics* 22:400–407.

Rochowiak, Dan. 2005. "Search and Heuristics." University of Alabama in Huntsville. Unpublished ms.

Roemer, John. 2001. *Political Competition: Theory and Applications*. Cambridge, MA: Harvard University Press.

Rothschild, Michael. 1974. "A Two-Armed Bandit Theory of Market Pricing." *Journal of Economic Theory* 9:185–202.

Rubinstein, Ariel. 1998. *Modeling Bounded Rationality*. Cambridge, MA: MIT Press.

Sagan, Scott. 1993. *The Limits of Safety: Organizations, Accidents, and Nuclear Weapons*. Princeton, NJ: Princeton University Press.

Samuels, Richard, and Stephen Stich. 2004. "Rationality and Psychology." In A. Mele and P. Rawling (eds.), *The Oxford Handbook of Rationality*, pp. 279–300. New York: Oxford University Press.

Samuels, Richard, Stephen Stich, and Michael Bishop. 2002. "Ending the Rationality Wars: How to Make Disputes about Human Rationality Disappear." In R. Elio (ed.), *Common Sense, Reasoning, and Rationality* 11:236–68. New York: Oxford University Press.

Samuels, Richard, Stephen Stich, and Luc Faucher. 2004. "Reason and Rationality." In I. Niiniluoto, M. Sintonen, and J. Wolenski (eds.), *Handbook of Epistemology*, pp. 1–50. Dordrecht: Kluwer.

Sapolsky, Harvey. 1972. *The Polaris System Development: Bureaucratic and Programmatic Success in Government*. Cambridge, MA : Harvard University Press.

Schachter, Daniel. 2002. *The Seven Sins of Memory: How the Mind Forgets and Remembers*. New York: Houghton Mifflin.

Schenking, J. Charles. 2005. *Making Waves: Politics, Propaganda, and the Emergence of the Imperial Japanese Navy, 1868–1922*. Stanford, CA: Stanford University Press.

Scherer, F. M. 1970. *Industrial Market Structure and Economic Performance*. Chicago: Rand McNally.

Schotter, Andrew. 2006. "Strong and Wrong: The Use of Rational Choice Theory in Experimental Economics." *Journal of Theoretical Politics* 18(4):498–511.

Schulman, Paul. 1975. "Nonincremental Policy Making: Notes toward an Alternative Paradigm." *American Political Science Review* 69:1354–70.

Schwartz, Barr, Andrew Ward, John Monterosso, Sonja Lyubomirsky, Katherine White, and Darrin Lehman. 2002. "Maximizing Versus Satisficing: Happiness Is a Matter of Choice." *Journal of Personality and Social Psychology* 83(5):1178–97.

Scott, James. 1998. *Seeing Like a State: How Certain Schemes to Improve the Human Condition Have Failed.* New Haven, CT: Yale University Press.

Scott, W. Richard. 1981. *Organizations: Rational, Natural, and Open Systems.* 1st ed. Englewood Cliffs, NJ: Prentice-Hall.

Scott, W. Richard. 1992. *Organizations: Rational, Natural, and Open Systems.* 3rd ed. Englewood Cliffs, NJ: Prentice-Hall.

Selten, Reinhard. 1989. "Evolution, Learning, and Economic Behavior." *Games and Economic* Behavior 3:3–24.

Shafir, Eldar, and Amos Tversky. 2002. "Decision Making." In D. Levitin (ed.), *Foundations of Cognitive Psychology*, pp. 601–20. Cambridge, MA: MIT Press.

Shanteau, James, and Rickey Thomas. 2000. "Fast and Frugal Heuristics: What about Unfriendly Environments?" *Behavioral and Brain Sciences* 23:762–63.

Shapere, Dudley. 1984. *Reason and the Search for Knowledge: Investigations in the Philosophy of Science.* Dordrecht: D. Reidel Publishing Company.

Shepsle, Kenneth. 1996. "Statistical Political Philosophy and Positive Political Theory." In J. Friedman (ed.), *The Rational Choice Controversy.* New Haven, CT: Yale University Press.

Simon, Herbert. 1947. *Administrative Behavior.* New York: Free Press.

Simon, Herbert. 1955. "A Behavioral Model of Rational Choice." *Quarterly Journal of Economics* 69 (February): 99–118.

Simon, Herbert. 1956. "Rational Choice and the Structure of the Environment." *Psychological Review* 63:129–38.

Simon, Herbert. 1957. *Models of Man: Social and Rational.* New York: Wiley.

Simon, Herbert. 1962. "The Architecture of Complexity." Proceedings of the American Philosophical Society 106:467–82.

Simon, Herbert. 1964. "The Concept of Organizational Goal." *Administrative Science Quarterly* 9:1–22.

Simon, Herbert. 1978. "Rationality as Process and Product of Thought." *American Economic Review* 68(2):1–16.

Simon, Herbert. 1979a. *Models of Thought.* Vol. 1. New Haven, CT: Yale University Press.

Simon, Herbert. 1979b. "Rational Decision Making in Business Organizations." *American Economic Review* 69:493–513.

Simon, Herbert. 1987. "Politics as Information-Processing." *London School of Economics Quarterly* 1:345–70.

Simon, Herbert. 1990. "Invariants of Human Behavior." In M. Rosenzweig and L. Porter (eds.), *Annual Review of Psychology* 41:1–19. Palo Alto: Annual Reviews.

Simon, Herbert. 1991. *Models of My Life.* New York: Basic Books.

Simon, Herbert. 1996. *The Sciences of the Artificial.* 3rd ed. Cambridge, MA: MIT Press.

Simon, Herbert. 1999a. "The Potlatch between Economics and Political Science." In J. Alt, M. Levi, and E. Ostrom (eds.), *Competition and Cooperation: Conversations with Nobelists about Economics and Political Science.* New York: Russell Sage.

Simon, Herbert. 1999b. "Problem Solving." In R. Wilson and F. Keil (eds.). *The MIT Encyclopedia of the Cognitive Sciences,* pp. 674–76. Cambridge, MA: MIT Press.

Simon, Herbert, and William Chase. 1973. "Skill in Chess." *American Scientist* 61:394–403.

Simon, Herbert, and J. Schaeffer. 1992. "The Game of Chess." In R. Aumann and S. Hart (eds.), *Handbook of Game Theory with Economic Applications,* 1:1–17. Amsterdam: North-Holland.

Simon, Herbert, and Peter Simon. 1962. "Trial and Error Search in Solving Difficult Problems: Evidence from the Game of Chess." *Behavioral Science* 7:425–29.

Simon, Herbert, Donald Smithburg, and Victor Thompson. 1950. *Public Administration.* New York: Knopf.

Sjoblom, Gunnar. 1993. "Some Critical Remarks on March and Olsen's *Rediscovering Institutions.*" *Journal of Theoretical Politics* 5(3):397–407.

Slovic, Paul. 1990. "Choice." In E. Smith and D. Osherson (eds.), *Thinking: An Invitation to Cognitive Science* 3:89–116. Cambridge, MA: MIT Press

Slovic, Paul. 1995. "The Construction of Preference." *American Psychologist* 50:364–71.

Sniderman, Paul. 2000. "Taking Sides: A Fixed Choice Theory of Political Reasoning." In A. Lupia, M. McCubbins, and S. Popkin (eds.), *Elements of Reason,* pp. 67–84. Cambridge: Cambridge University Press.

Sniderman, Paul, Richard Brody, and Philip Tetlock. 1991. *Reasoning and Choice: Explorations in Political Psychology.* New York: Cambridge University Press.

Speer, Albert. 1970. *Inside the Third Reich.* New York: Macmillan.

Sproull, Lee, Stephen Weiner, and David Wolf. 1978. *Organizing an Anarchy.* Chicago: University of Chicago Press.

Stanovich, Keith. 1999. *Who Is Rational? Studies of Individual Differences in Reasoning.* Mahwah, NJ: Lawrence Erlbaum.

Stanovich, Keith, and Richard West. 2000. "Individual Differences in Reasoning: Implications for the Rationality Debate." *Behavioral and Brain Sciences* 23:645–65.

Stinchcombe, Arthur. 1965. "Social Structure and Organizations." In J. March (ed.), *Handbook of Organizations,* pp. 142–93. Chicago: Rand McNally.

Stokes, Susan. 1999. "Political Parties and Democracy." *Annual Review of Political Science* 2:243–67.

Sugihara, Kaoru. 2004. "The State and the Industrious Revolution in Tokugawa Japan." Working Paper No. 02/04. Graduate School of Economics, Osaka University.

Sunstein, Cass, Reid Hastie, John Payne, David Schkade, and W. Kip Vicusi. 2002. *Punitive Damages: How Juries Decide.* Chicago: University of Chicago Press.

Surowiecki, James. 2005. *The Wisdom of Crowds.* New York: Anchor Books.

Taylor, Frederick. [1911] 1947. *The Principles of Scientific Management.* New York: Harper & Row.

Taylor, Serge. 1984. *Making Bureaucracies Think*. Stanford, CA: Stanford University Press.

Tetlock, Philip. 1999. "Theory-Driven Reasoning about Possible Pasts and Probable Futures in World Politics: Are We Prisoners of Our Preconceptions?" *American Journal of Political Science* 43:335–66.

Tetlock, Philip. 2005. *Expert Political Judgment: How Good Is It? How Can We Know?* Princeton, NJ: Princeton University Press.

Thompson, James. 1980. *Rolling Thunder: Understanding Policy and Program Failure*. Chapel Hill: University of North Carolina Press.

Ting, Michael. 2003. "A Strategic Theory of Bureaucratic Redundancy." *American Journal of Political Science* 47:274–92.

Todd, Peter, and Gerd Gigerenzer. 2000. "Authors' Response." *Behavioral and Brain Sciences* 23:767–77.

Triandis, Harry. 1995. *Individualism and Collectivism*. Boulder, CO: Westview Press.

Tversky, Amos, and Daniel Kahneman. 1974. "Judgments under Uncertainty: Heuristics and Biases." *Science* 185:1124–31.

Tversky, Amos, and Daniel Kahneman. 1986. "Rational Choice and the Framing of Decisions." *Journal of Business* 59:S251–78.

Udy, Stanley. 1959. "'Bureaucracy' and 'Rationality' in Weber's Organization Theory." *American Sociological Review* 24:791–95.

Udy, Stanley. 1962. "Administrative Rationality, Social Setting, and Organizational Development." *American Journal of Sociology* 68:299–308.

Udy, Stanley. 1965. "The Comparative Analysis of Organizations." In J. March (ed.), *Handbook of Organizations,* pp. 678–709. Chicago: Rand McNally.

Van der Panne, Gerben, Cees van Beers, and Alfred Kleinknecht. 2003. "Success and Failure of Innovation: A Literature Review." *International Journal of Innovation Management* 7(3):1–30.

Von Winterfeldt, Detlef, and Ward Edwards. 1986. *Decision Analysis and Behavioral Research*. Cambridge: Cambridge University Press.

Wallin, Annika, and Peter Gardenfors. 2000. "Smart People Who Make Simple Heuristics Work." *Behavioral and Brain Sciences* 23:765.

Watson, James D. 1968. *The Double Helix*. New York: Atheneum.

Weick, Karl. 1979. *The Social Psychology of Organizing*. 2nd ed. Reading, MA: Addison-Wesley.

Wegner, Daniel, and John Bargh. 1998. "Control and Automaticity in Social Life." In D. T. Gilbert, S. T. Fiske, and G. Lindzey (eds.), *Handbook of Social Psychology,* pp. 446–95. 4th ed. New York: McGraw-Hill.

Weiner, Stephen. 1976. "Participation, Deadlines, and Choice." In March and Olsen (eds.), *Ambiguity and Choice in Organizations,* pp. 225–50.

Weingast, Barry. 1996. "Political Institutions: Rational Choice Perspectives." In R. E. Goodin and H. Klingemann (eds.), *A New Handbook of Political Science,* pp. 167–190. Oxford: Oxford University Press.

Weissinger-Baylon, Roger. 1986. "Garbage Can Decision Processes in Naval Warfare." In March and Weissinger-Baylon (eds.), *Ambiguity and Command,* pp. 36–52.

Whittle, Peter. 1982. *Optimization over Time: Dynamic Programming and Stochastic Control*. New York: Wiley.

Wildavsky, Aaron. 1964. *The Politics of the Budgetary Process*. Boston: Little, Brown.

Wildavsky, Aaron. 1972. "The Self-Evaluating Organization." *Public Administration Review* 32(5):509–20.

Wildavsky, Aaron. 1979. *Speaking Truth to Power: The Art and Craft of Policy Analysis*. Boston: Little, Brown.

Wilensky, Harold. 1967. *Organizational Intelligence*. New York: Basic Books.

Williamson, Samuel. 1979. "Theories of Organizational Process and Foreign Policy Outcomes." In P. Lauren (ed.), *Diplomacy: New Approaches in History, Theory, and Policy*, pp. 137–61. New York: Free Press.

Wilson, James Q. 1989. *Bureaucracy: What Government Agencies Do and Why They Do It*. New York: Basic Books.

Woodward, David. 1965. *The Russians at Sea*. London: William Kimber.

# Index

Text:        10/13 Sabon
Display:     Sabon
Compositor:  Publication Services, Inc.

Milton Keynes UK
Ingram Content Group UK Ltd.
UKHW040139141024
449609UK00005B/125